Clang Compiler Frontend

Get to grips with the internals of a C/C++ compiler
frontend and create your own tools

Ivan Murashko

Clang Compiler Frontend

Associate Group Product Manager: Kunal Sawant
Senior Editor: Rounak Kulkarni
Senior Content Development Editor: Rosal Colaco
Technical Editor: Jubit Pincy
Copy Editor: Safis Editing
Project Coordinator: Deeksha Thakkar
Indexer: Pratik Shirodkar
Production Designer: Vijay Kamble
Business Development Executive: Debadrita Chatterjee
Senior Developer Relations Marketing Executive: Shrinidhi Monaharan

First published: March 2024

Production reference: 1290224

Published by Packt Publishing Ltd.
Grosvenor House
11 St Paul's Square
Birmingham
B3 1RB, UK

ISBN 978-1-83763-098-1

www.packtpub.com

Contributors

About the author

Ivan Murashko is a C++ software engineer. He earned his Ph.D. in Physics from Peter the Great St. Petersburg Polytechnic University and has over 20 years of C++ programming experience, mostly on Linux. Since 2020, he has worked with LLVM compilers and has been an LLVM committer since 2021. His areas of interest include the Clang compiler frontend, Clang Tools (such as Clang-Tidy and Clangd), and performance optimizations for compilers and compiler tools.

I want to thank my wife, Irina, who was patient and supported me throughout the writing of this book.

About the reviewer

Aditya Agrawal comes from the city of joy, Kolkata, West Bengal. He is currently working as a Software Engineer in the Systems Domain. He graduated with a master's degree in Computer Science from the reputed Indian Institute of Technology, Madras where he was introduced to the world of Compilers, Parallel Programming, and systems. Aditya has published a research paper that allows one to add point-to-point synchronizations for their OpenMP parallel programs (called UWPro). He has read a lot of books and tutorials on how to work with Compilers and the like. Aditya has experience working with RISCV during his tenure at MIPS Embedded Technologies as a full-time RISCV Developer. In his spare time, he loves to play video games and take part in various community events involving LLVM Social, RV Bangalore User Group, and so on.

Table of Contents

Preface

Low Level Virtual Machine (LLVM), is a collection of modular and reusable compiler and toolchain technologies used to develop compilers and compiler tools, such as linters and refactoring tools. LLVM is written in C++ and can be considered a good example of a well-structured project that uses interesting techniques aimed at making it reusable and efficient. The project can also be considered an excellent example of compiler architecture; diving into it will give you a sense of how compilers are organized and how they function. This should help to understand usage patterns and apply them accordingly.

One of the key components of LLVM is the C/C++ compiler known as Clang. This compiler is widely used across various companies and has been designated as the default compiler for certain development environments, notably for macOS development. Clang will be the primary focus of our investigation in this book, with particular attention to its frontend—the part that is closest to the C/C++ programming language. Specifically, the book will include examples demonstrating how the C++ standard is implemented within the compiler.

A pivotal aspect of LLVM's design is its modularity, which facilitates the creation of custom tools that exploit the compiler's comprehensive capabilities. A notable example covered in the book is the Clang-Tidy linter framework, designed to identify undesirable code patterns and recommend corrections. Although it includes several hundred checks, you may not find one specific to your project's needs. However, the book will provide you with the foundation necessary to develop such a check from the beginning.

LLVM is an actively evolving project with two major releases each year. At the time the book was written, the latest stable release was version 17. Meanwhile, a release candidate for version 18 was introduced in January 2024, with its official release anticipated to coincide with the publication of the book. The book's content has been verified against the latest compiler version, 18, ensuring it provides insights based on the most current compiler implementation available.

Who this book is for

The book is written for C++ engineers who don't have prior knowledge of compilers but wish to gain this knowledge and apply it to their daily activities. It provides an overview of the Clang compiler frontend, an essential yet often underestimated part of LLVM. This section of the compiler, along with a collection of powerful tools, enables programmers to enhance code quality and the overall development process. For example, Clang-Tidy offers more than 500 different lint checks that detect anti-patterns in code (such as use after move) and help maintain code style and standards. Another notable tool is Clang-Format, which allows specifying various formatting rules suitable for your project. These tools can also be considered an integral part of the development process. For instance, the language server (Clangd) is a critical service providing navigation and refactoring support for your IDE.

Understanding compiler internals might be crucial for anyone wanting to create and use such tools. The book provides the necessary foundation to begin this journey, covering basic LLVM architecture and offering a detailed description of Clang internals. It includes examples from LLVM source code and custom tools that extend the basic functionality provided by the compiler. Additionally, the book addresses compilation databases and various performance optimizations that can enhance the build speed of your projects. This knowledge should help C++ developers correctly apply the compiler to their work activities.

What this book covers

Chapter 1, Environment Setup, describes the basic steps required to set up the environment for future experiments with Clang, suitable for Unix-based systems such as Linux and

Darwin (macOS). In addition, readers will learn how to download, configure, and build LLVM source code. We will also create a simple Clang Tool to verify the syntax of the provided source code.

Chapter 2, Clang Architecture, examines the internal architecture of the Clang compiler. Starting with the basic concept of a compiler, we will explore how it is implemented in Clang. We will look at various parts of the compiler, including the driver, preprocessor (lexer), and parser. We will also examine examples that show how the C++ standard is implemented in Clang.

Chapter 3, Clang AST, talks about **Clang Abstract Syntax Tree (AST)**, which is the basic data structure produced by the parser. We will explore how the AST is organized in Clang and how it can be traversed. We will also delve into AST Matchers — a powerful tool provided by Clang for locating specific AST nodes.

Chapter 4, Basic Libraries and Tools, explores basic LLVM libraries and tools, including the LLVM **Abstract Data Type (ADT)** library, used across all LLVM code. We will investigate TableGen, a **Domain-Specific Language (DSL)** used to generate C++ code in various parts of LLVM. Additionally, we will explore **LLVM Integrated Tester (LIT)** tool used for creating powerful end-to-end tests. Using the knowledge gained, we will create a simple Clang plugin to estimate source code complexity.

Chapter 5, Clang-Tidy Linter Framework, covers Clang-Tidy, a linter framework based on Clang AST, and creates a simple Clang-Tidy check. We will also discuss how compilation errors affect the AST and the results provided by different Clang Tools, such as Clang-Tidy.

Chapter 6, Advanced Code Analysis, goes further and considers another advanced data structure used for code analysis: **Control Flow Graphs (CFG)**. We will investigate typical cases for its application and create a simple Clang-Tidy check that utilizes this data structure.

Chapter 7, Refactoring Tools, Clang provides advanced tools for code modification and refactoring. We will explore different ways to create a custom refactoring tool, including one based on the Clang-Tidy linter framework. We will also explore Clang-Format, an extremely fast utility for automatic code formatting.

Chapter 8, IDE Support and Clangd, presents Clangd - a Language Server used in various IDEs, such as **Visual Studio Code (VS Code),** to provide intelligent support, including navigation and code modification. Clangd exemplifies the utility of the powerful modular architecture of LLVM. It utilizes various Clang tools, such as Clang-Tidy and Clang-Format, to enhance the development experience in VS Code. Compiler performance is crucial for this tool, and we will explore several techniques Clangd employs to improve its performance, thereby offering the best experience to developers.

Appendix 1: Compilation Database, describes the Compilation Database—a method for providing complex compilation commands to different Clang Tools. This functionality is crucial for integrating Clang Tools such as Clangd and Clang-Tidy into real C/C++ projects.

Appendix 2: Build Speed Optimizations, covers several compiler performance optimizations that can be used to enhance compiler performance. We will cover Clang precompiled headers and Clang modules, which represent a serialized AST that can be loaded much faster than building it from scratch.

To get the most out of this book

You will need to have an understanding of C++, especially C++17, which is used for LLVM and throughout the examples in the book. The provided examples are assumed to be run on a Unix-like operating system, with Linux and Darwin (Mac OS) being considered the operating system requirements for the book. We will use Git to clone the LLVM source tree and start working on it. Some tools also need to be installed, such as CMake and Ninja, which will be actively used to build the examples and the LLVM source code.

If you are using the digital version of this book, we advise you to type the code yourself or access the code from the book's GitHub repository (a link is available in the next section). Doing so will help you avoid any potential errors related to the copying and pasting of code.

Download the example code files

The code bundle for the book is also hosted on GitHub at `https://github.com/PacktPu
blishing/Clang-Compiler-Frontend-Packt`. In case there's an update to the code, it will
be updated on the existing GitHub repository.

We also have other code bundles from our rich catalog of books and videos available at
`https://github.com/PacktPublishing/`. Check them out!

Conventions used

There are a number of text conventions used throughout this book.

`CodeInText`: Indicates code words in text, database table names, folder names, filenames,
file extensions, pathnames, dummy URLs, and user input. Here is an example: "The first two
parameters specify the declaration (`clang::Decl`) and the statement for the declaration
(`clang::Stmt`)."

A block of code is set as follows:

```
1  int main() {
2    return 0;
3  }
```

Any command-line input or output is written as follows:

```
$ ninja clang
```

We use `<...>` as a placeholder for the folder where the LLVM source code was cloned.

Some code examples will be representing input of shells. You can recognize them by specific
prompt characters:

- **(lldb)** for interactive LLDB shell
- $ for Bash shell (macOS and Linux)
- > for interactive shell provided by different Clang Tools, such as Clang-Query

> **Important note**
>
> Warnings or important notes appear like this.

> **Tip**
>
> Tips and tricks appear like this.

Get in touch

Feedback from our readers is always welcome.

General feedback: If you have questions about any aspect of this book, mention the book title in the subject of your message and email us at customercare@packtpub.com.

Errata: Although we have taken every care to ensure the accuracy of our content, mistakes do happen. If you have found a mistake in this book, we would be grateful if you would report this to us. Please visit `https://www.packtpub.com/support/errata`, selecting your book, clicking on the Errata Submission Form link, and entering the details.

Piracy: If you come across any illegal copies of our works in any form on the Internet, we would be grateful if you would provide us with the location address or website name. Please contact us at copyright@packt.com with a link to the material.

If you are interested in becoming an author: If there is a topic that you have expertise in and you are interested in either writing or contributing to a book, please visit `https://partnerships.packt.com/contributors/`.

Share your thoughts

Once you've read *Clang Compiler Frontend*, we'd love to hear your thoughts! Scan the QR code below to go straight to the Amazon review page for this book and share your feedback.

https://packt.link/r/1837630984

Your review is important to us and the tech community and will help us make sure we're delivering excellent quality content.

Download a free PDF copy of this book

Thanks for purchasing this book!

Do you like to read on the go but are unable to carry your print books everywhere? Is your eBook purchase not compatible with the device of your choice?

Don't worry, now with every Packt book you get a DRM-free PDF version of that book at no cost.

Read anywhere, any place, on any device. Search, copy, and paste code from your favorite technical books directly into your application.

The perks don't stop there, you can get exclusive access to discounts, newsletters, and great free content in your inbox daily.

Follow these simple steps to get the benefits:

1. Scan the QR code or visit the link below:

https://download.packt.com/free-ebook/9781837630981

2. Submit your proof of purchase.

3. That's it! We'll send your free PDF and other benefits to your email directly.

Part 1

Clang Setup and Architecture

You can find some info about LLVM internal architecture and how Clang fits into it. There is also description how to install and build Clang and Clang-Tools, description for basic LLVM libraries and tools used across LLVM project and essential for Clang development. You can find description for some Clang features and their internal implementation.

This part has the following chapters:

1

Environment Setup

In this chapter, we will discuss the basic steps of setting up the environment for future experiments with Clang . The setup is appropriate for Unix-based systems such as Linux and Mac OS (Darwin). In addition, you will get important information on how to download, configure, and build the LLVM source code. We will continue with a short session that explains how to build and use the **LLVM debugger (LLDB**), which will be used as the primary tool for code investigation throughout the book. Finally, we will finish with a simple Clang tool that can check C/C++ files for compilation errors. We will use LLDB for a simple debug session for the created tool and clang internal. We will cover the following topics:

- Prerequisites

- Getting to know LLVM

- Source code compilation

- How to create a custom Clang tool

1.1   Technical requirements

Downloading and building LLVM code is very easy and does not require any paid tools. You will require the following:

- Unix-based OS (Linux, Darwin)

- Command line git

- Build tools: CMake and Ninja

We will use the debugger as the source investigation tool. LLVM has its own debugger, LLDB. We will build it as our first tool from LLVM monorepo: `https://github.com/llvm/llvm-project.git`.

Any build process consists of two steps. The first one is the project configuration and the last one is the build itself. LLVM uses CMake as a project configuration tool. It also can use a wide range of build tools, such as Unix Makefiles, and Ninja. It can also generate project files for popular IDEs such as Visual Studio and XCode. We are going to use Ninja as the build tool because it speeds up the build process, and most LLVM developers use it. You can find additional information about the tools here: `https://llvm.org/docs/GettingStarted.html`.

The source code for this chapter is located in the `chapter1` folder of the book's GitHub repository: `https://github.com/PacktPublishing/Clang-Compiler-Frontend-Packt/tree/main/chapter1`

1.1.1   CMake as project configuration tool

CMake is an open source, cross-platform build system generator. It has been used as the primary build system for LLVM since version 3.3, which was released in 2013.

Before LLVM began using CMake, it used autoconf, a tool that generates a configure script that can be used to build and install software on a wide range of Unix-like systems. However, autoconf has several limitations, such as being difficult to use and maintain and having poor support for cross-platform builds. CMake was chosen as an alternative to autoconf because it addresses these limitations and is easier to use and maintain.

In addition to being used as the build system for LLVM, CMake is also used for many other software projects, including Qt, OpenCV, and Google Test.

1.1.2   Ninja as build tool

Ninja is a small build system with a focus on speed. It is designed to be used in conjunction with a build generator, such as CMake, which generates a build file that describes the build rules for a project.

One of the main advantages of Ninja is its speed. It is able to execute builds much faster than other build systems, such as Unix Makefiles, by only rebuilding the minimum set of files necessary to complete the build. This is because it keeps track of the dependencies between build targets and only rebuilds targets that are out of date.

Additionally, Ninja is simple and easy to use. It has a small and straightforward command-line interface, and the build files it uses are simple text files that are easy to read and understand.

Overall, Ninja is a good choice for build systems when speed is a concern, and when a simple and easy-to-use tool is desired.

One of the most useful Ninja option is -j . This option allows you to specify the number of threads to be run in parallel. You may want to specify the number depending on the hardware you are using.

Our next goal is to download the LLVM code and investigate the project structure. We also need to set up the necessary utilities for the build process and establish the environment for our future experiments with LLVM code. This will ensure that we have the tools and dependencies in place to proceed with our work efficiently.

1.2   Getting to know LLVM

Let's begin by covering some foundational information about LLVM, including the project history as well as its structure.

1.2.1   Short LLVM history

The Clang compiler is a part of the LLVM project. The project was started in 2000 by Chris Lattner and Vikram Adve as their project at the University of Illinois at Urbana–Champaign [26].

LLVM was originally designed to be a next-generation code generation infrastructure that could be used to build optimizing compilers for many programming languages. However, it has since evolved into a full-featured platform that can be used to build a wide variety of tools, including debuggers, profilers, and static analysis tools.

LLVM has been widely adopted in the software industry and is used by many companies and organizations to build a variety of tools and applications. It is also used in academic research and teaching and has inspired the development of similar projects in other fields.

The project received an additional boost when Apple hired Chris Lattner in 2005 and formed a team to work on LLVM. LLVM became an integral part of the development tools created by Apple (XCode).

Initially, **GNU Compile Collection (GCC)** was used as the C/C++ frontend for LLVM. But that had some problems. One of them was related to GNU **General Public License (GPL)** that prevented the frontend usage at some proprietary projects. Another disadvantage was the limited support for Objective-C in GCC at the time, which was important for Apple. The Clang project was started by Chris Lattner in 2006 to address the issues.

Clang was originally designed as a unified parser for the C family of languages, including C, Objective-C, C++, and Objective-C++. This unification was intended to simplify maintenance by using a single frontend implementation for multiple languages, rather than maintaining multiple implementations for each language. The project became successful very quickly. One of the primary reasons for the success of Clang and LLVM was their modularity. Everything in LLVM is a library, including Clang . It opened the opportunity to create a lot of amazing tools based on Clang and LLVM, such as clang-tidy and clangd, which will be covered later in the book (*Chapter 5, Clang-Tidy Linter Framework* and *Chapter 8, IDE Support and Clangd*).

LLVM and Clang have a very clear architecture and are written in C++. That makes it possible to be investigated and used by any C++ developer. We can see the huge community created around LLVM and the extremely fast growth of its usage.

1.2.2 OS support

We are planning to focus on OS for personal computers here, such as Linux, Darwin, and Windows. On the other hand, Clang is not limited by personal computers but can also be used to compile code for mobile platforms such as iOS and different embedded systems.

Linux

The GCC is the default set of dev tools on Linux, especially gcc (for C programs) and g++ (for C++ programs) being the default compilers. Clang can also be used to compile source code on Linux. Moreover, it mimics to gcc and supports most of its options. LLVM support might be limited for some GNU tools, however; for instance, GNU Emacs does not support LLDB as a debugger. But despite this, Linux is the most suitable OS for LLVM development and investigation, thus we will mainly use this OS (Fedora 39) for future examples.

Darwin (macOS)

Clang is considered the main build tool for Darwin. The entire build infrastructure is based on LLVM, and Clang is the default C/C++ compiler. The developer tools, such as the debugger (LLDB), also come from LLVM. You can get the primary developer utilities from XCode, which are based on LLVM. However, you may need to install additional command-line tools, such as CMake and Ninja, either as separate packages or through package systems such as MacPorts or Homebrew.

For example, you can get CMake using Homebrew as follows:

```
$ brew install cmake
```

or for MacPorts:

```
$ sudo port install cmake
```

Windows

On Windows, Clang can be used as a command-line compiler or as part of a larger development environment such as Visual Studio. Clang on Windows includes support for the **Microsoft Visual C++ (MSVC)** ABI, so you can use Clang to compile programs that use the **Microsoft C runtime library (CRT)** and the C++ **Standard Template Library (STL)**. Clang also supports many of the same language features as GCC, so it can be used as a drop-in replacement for GCC on Windows in many cases.

It's worth mentioning clang-cl [9]. It is a command-line compiler driver for Clang that is designed to be used as a drop-in replacement for the MSVC compiler, cl.exe . It was introduced as part of the Clang compiler, and is created to be used with the LLVM toolchain.

Like cl.exe , clang-cl is designed to be used as part of the build process for Windows programs, and it supports many of the same command-line options as the MSVC compiler. It can be used to compile C, C++, and Objective-C code on Windows, and it can also be used to link object files and libraries to create executable programs or **dynamic link libraries (DLLs)**.

The development process for Windows is different from that of Unix-like systems, which require additional specifics that might make the book material quite complicated. To avoid this complexity, our primary goal is to focus on Unix-based systems, such as Linux and Darwin, and we will omit Windows-specific examples in this book.

1.2.3   LLVM/Clang project structure

The Clang source is a part of the LLVM **monolithic repository (monorepo)**. LLVM started to use the monorepo in 2019 as a part of its transition to Git [4]. The decision was driven by several factors, such as better code reuse, improved efficiency, and collaboration. Thus you can find all the LLVM projects in one place. As mentioned in the Preface, we will be using LLVM version 18.x in this book. The following command will allow you to download it:

```
$ git clone https://github.com/llvm/llvm-project.git -b release/18.x
$ cd llvm-project
```

Figure 1.1: Getting the LLVM code base

The most important parts of the **llvm-project** that will be used in the book are shown in Figure 1.2.

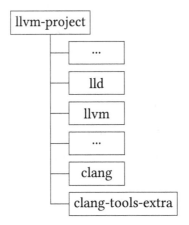

Figure 1.2: LLVM project tree

There are:

- lld : The LLVM linker tool. You may want to use it as a replacement for standard linker tools, such as GNU ld

- `llvm` : Common libraries for LLVM projects

- `clang` : The clang driver and frontend

- `clang-tools-extra` : These are different clang tools that will be covered in the second part of the book

Most projects have the structure shown in Figure 1.3.

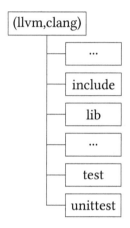

Figure 1.3: Typical LLVM project structure

LLVM projects, such as `clang` or `llvm`, typically contain two primary folders: `include` and `lib`. The `include` folder contains the project interfaces (header files), while the `lib` folder contains the implementation. Each LLVM project has a variety of different tests, which can be divided into two primary groups: unit tests located in the `unittests` folder and implemented using the Google Test framework, and end-to-end tests implemented using the **LLVM Integrated Tester (LIT)** framework. You can get more info about LLVM/Clang testing in *Section 4.5.2, LLVM test framework*.

The most important projects for us are `clang` and `clang-tools-extra` . The `clang` folder contains the frontend and driver.

> **Important note**
>
> The compiler driver is used to run different stages of compilation (parsing, optimization, link, and so on.). You can get more info about it at *Section 2.3, Clang driver overview*.

For instance, the lexer implementation is located in the `clang/lib/Lex` folder. You can also see the `clang/test` folder, which contains end-to-end tests, and the `clang/unittest` folder, which contains unit tests for the frontend and the driver.

Another important folder is `clang-tools-extra`. It contains some tools based on different Clang libraries. They are as follows:

- `clang-tools-extra/clangd` : A language server that provides navigation info for IDEs such as VSCode

- `clang-tools-extra/clang-tidy` : A powerful lint framework with several hundred different checks

- `clang-tools-extra/clang-format` : A code formatting tool

After obtaining the source code and setting up build tools, we are ready to compile the LLVM source code.

1.3   Source code compilation

We are compiling our source code in debug mode to make it suitable for future investigations with a debugger. We are using LLDB as the debugger. We will start with an overview of the build process and finish building the LLDB as a concrete example.

1.3.1   Configuration with CMake

Create a build folder where the compiler and related tools will be built:

```
$ mkdir build
$ cd build
```

The minimal configuration command looks like this:

```
$ cmake -DCMAKE_BUILD_TYPE=Debug ../llvm
```

The command requires the build type to be specified (e.g. `Debug` in our case) as well as the primary argument that points to a folder with the build configuration file. The configuration file is stored as `CMakeLists.txt` and is located in the `llvm` folder, which explains the `../llvm` argument usage. The command generates `Makefile` located in the build folder, thus you can use the simple `make` command to start the build process.

We will use more advanced configuration commands in the book. One of the commands looks like this:

```
cmake -G Ninja -DCMAKE_BUILD_TYPE=Debug -DCMAKE_INSTALL_PREFIX=../install
↪ -DLLVM_TARGETS_TO_BUILD="X86"
↪ -DLLVM_ENABLE_PROJECTS="lldb;clang;clang-tools-extra"
↪ -DLLVM_USE_SPLIT_DWARF=ON ../llvm
```

Figure 1.4: Basic CMake configuration

The are several LLVM/cmake options specified:

- `-G Ninja` specifies Ninja as the build generator, otherwise it will use make (which is slow).

- `-DCMAKE_BUILD_TYPE=Debug` sets the build mode. The build with debug info will be created. There is a primary build configuration for Clang internals investigations.

- `-DCMAKE_INSTALL_PREFIX=../install` specifies the installation folder.

- `-DLLVM_TARGETS_TO_BUILD="X86"` sets exact targets to be build. It will avoid building unnecessary targets.

- `-DLLVM_ENABLE_PROJECTS="lldb;clang;clang-tools-extra"` specifies the LLVM projects we want to build.

- `-DLLVM_USE_SPLIT_DWARF=ON` splits debug information into separate files. This option saves disk space as well as memory consumption during the LLVM build.

We used `-DLLVM_USE_SPLIT_DWARF=ON` to save some space on the disk. For instance, the Clang build (`ninja clang` build command) with the option enabled takes up 20 GB, but it takes up 27 GB space with the option disabled. Note that the option requires a compiler used for the build to support it. You might also notice that we create the build for one specific architecture: `X86` . This option also saved some space for us because otherwise, all supported architecture will be built and the required space will also increase from 20 GB to 27 GB.

> **Important note**
>
> You might want to avoid using the `-DLLVM_TARGETS_TO_BUILD="X86"` setting if your host platform is different from X86, for instance, ARM. For ARM, you can use the following configuration: `-DLLVM_TARGETS_TO_BUILD="ARM;X86;AArch64"` [15]. The full list of supported platforms can be found in [7] and includes (as of March 2023) 19 different targets.
>
> You can also use the default settings and not specify the `LLVM_TARGETS_TO_BUILD` configuration setting. Be prepared for both an increase in build time and the amount of space used.

You can save more space if you use dynamic libraries instead of static ones. The configuration setting `-DBUILD_SHARED_LIBS=ON` will build each LLVM component as a shared library. The space used will be 14 GB, and the overall config command will look like this:

```
cmake -G Ninja -DCMAKE_BUILD_TYPE=Debug -DCMAKE_INSTALL_PREFIX=../install
 ↪  -DLLVM_TARGETS_TO_BUILD="X86"
 ↪  -DLLVM_ENABLE_PROJECTS="lldb;clang;clang-tools-extra"
 ↪  -DLLVM_USE_SPLIT_DWARF=ON -DBUILD_SHARED_LIBS=ON ../llvm
```

Figure 1.5: CMake configuration that enables shared libraries instead of static ones

For performance purposes, on Linux, you might want to use the `gold` linker instead of the default one. The `gold` linker is an alternative to the GNU Linker, which was developed as part of the GNU Binary Utilities (binutils) package. It is designed to be faster and more efficient than the GNU Linker, especially when linking large projects. One way it achieves this is by using a more efficient algorithm for symbol resolution and a more compact file format for the resulting executable. It can be enabled with the -DLLVM_USE_LINKER=gold option. The result configuration command will look like this:

```
cmake -G Ninja -DCMAKE_BUILD_TYPE=Debug -DCMAKE_INSTALL_PREFIX=../install
 ↪  -DLLVM_TARGETS_TO_BUILD="X86"
 ↪  -DLLVM_ENABLE_PROJECTS="lldb;clang;clang-tools-extra"
 ↪  -DLLVM_USE_LINKER=gold -DLLVM_USE_SPLIT_DWARF=ON -DBUILD_SHARED_LIBS=ON
 ↪  ../llvm
```

Figure 1.6: CMake configuration that uses gold linker

The debug build can be very slow, so you may want to consider an alternative. A good compromise between debuggability and performance is the release build with debug information. To obtain this build, you can change the CMAKE_BUILD_TYPE flag to RelWithDebInfo in your overall configuration command. The command will then look like this:

```
cmake -G Ninja -DCMAKE_BUILD_TYPE=RelWithDebInfo
 ↪  _DCMAKE_INSTALL_PREFIX=../install -DLLVM_TARGETS_TO_BUILD="X86"
 ↪  -DLLVM_ENABLE_PROJECTS="lldb;clang;clang-tools-extra"
 ↪  -DLLVM_USE_SPLIT_DWARF=ON ../llvm
```

Figure 1.7: CMake configuration that uses RelWithDebInfo build type

The following table keeps the list of some popular options (`https://llvm.org/docs/CMake.html`).

Option	Description
CMAKE_BUILD_TYPE	Specifies the build configuration.
	Possible values are `Release|Debug|RelWithDebInfo|MinSizeRel`.
	`Release` and `RelWithDebInfo` are optimized for performance, while
	`MinSizeRel` is optimized for size.
CMAKE_INSTALL_PREFIX	Installation prefix
CMAKE_C,CXX_FLAGS	Extra C/C++ flags be used for compilation
CMAKE_C,CXX_COMPILER	C/C++ compiler be used for compilation.
	You might want to specify a non-default compiler to use some
	options that are not available or not supported by the default compiler.
LLVM_ENABLE_PROJECTS	The projects to be enabled. We will use `clang;clang-tools-extra`.
LLVM_USE_LINKER	Specifies the linker to be used.
	There are several options, including `gold` and `lld`.

Table 1.1: Configuration options

1.3.2 Build

We need to call Ninja to build the projects. If you want to build all specified projects, you can run Ninja without arguments:

```
$ ninja
```

The command for the Clang build will look like this:

```
$ ninja clang
```

You can also run unit and end-to-end tests for the compiler with the following:

```
$ ninja check-clang
```

The compiler binary is `bin/clang` and can be found in the `build` folder.

You can also install the binaries into the folder specified in the `-DCMAKE_INSTALL_PREFIX` option. It can be done as follows:

```
$ ninja install
```

The `../install` folder (specified as the installation folder in Figure 1.4) will have the following structure:

```
$ ls ../install
bin  include  lib  libexec  share
```

1.3.3   The LLVM debugger, its build, and usage

The LLVM debugger, LLDB , has been created with a look at the GNU debugger (GDB). Some of its commands repeat the counterparts from GDB . You may ask "Why do we need a new debugger if we have a good one?" The answer can be found in the different architecture solutions used by GCC and LLVM. LLVM uses a modular architecture, and different parts of the compiler can be reused. For example, the Clang frontend can be reused in the debugger, resulting in support for modern C/C++ features. For example, the print command in `lldb` can specify any valid language constructions, and you can use some modern C++ features with the `lldb` print command.

In contrast, GCC uses a monolithic architecture, and it's hard to separate the C/C++ frontend from other parts. Therefore, GDB has to implement language features separately, which may take some time before modern language features implemented in GCC become available in GDB .

You may find some info about LLDB build and a typical usage scenario in the following example. We are going to create a separate folder for the release build:

```
$ cd llvm-project
$ mkdir release
$ cd release
```

Figure 1.8: Release build for LLVM

We configure our project in release mode and specify the `lldb` and `clang` projects only:

```
cmake -G Ninja -DCMAKE_BUILD_TYPE=Release -DCMAKE_INSTALL_PREFIX=../install
 ↳  -DLLVM_TARGETS_TO_BUILD="X86" -DLLVM_ENABLE_PROJECTS="lldb;clang"
 ↳  ../llvm
```

Figure 1.9: CMake configuration that uses Release build type

We are going to build both Clang and LLDB using the maximum threads available in the system:

```
$ ninja clang lldb -j $(nproc)
```

You can install the created executables with the following command:

```
$ ninja install-clang install-lldb
```

The binary will be installed into the folder specified via the `-DCMAKE_INSTALL_PREFIX` config command argument.

We will use the following simple C++ program for the example debugger session:

```
1 int main() {
2   return 0;
3 }
```

Figure 1.10: Test C++ program: main.cpp

The program can be compiled using the following command (`<...>` was used to refer the folder where llvm-project was cloned):

```
$ <...>/llvm-project/install/bin/clang main.cpp -o main -g -O0
```

As you may have noticed, we don't use optimization (the `-O0` option) and store debug info in the binary (with the `-g` option).

A typical debug session for the created executable is shown in Figure 1.11.

```
1  $ <...>/llvm-project/install/bin/lldb main
2  (lldb) target create "./main"
3  ...
4  (lldb) b main
5  Breakpoint 1: where = main`main + 11 at main.cpp:2:3,...
6  (lldb) r
7  Process 1443051 launched: ...
8  Process 1443051 stopped
9  * thread #1, name = 'main', stop reason = breakpoint 1.1
10     frame #0: 0x000055555555513b main`main at main.cpp:2:3
11     1    int main() {
12  -> 2       return 0;
13     3    }
14  (lldb) q
```

Figure 1.11: LLDB session example

Several actions are taken:

- Run the debug session with `<...>/llvm-project/install/bin/lldb main`, where `main` is the executable we want to debug. See Figure 1.11, *Line 1.*

- We set a breakpoint in the `main` function. See Figure 1.11, *Line 4.*

- Run the session with "`r`" command. See Figure 1.11, *Line 6.*

- We can see that the process is interrupted at the breakpoint. See Figure 1.11, *Lines 8, 12.*

- We finish the session with the "`q`" command. See Figure 1.11, *Line 14.*

We are going to use LLDB as one of our tools for the Clang internal investigation. We will use the same sequence of commands that is shown in Figure 1.11. You can also use another debugger, such as GDB , that has a similar set of commands as LLDB .

1.4 Test project – syntax check with a Clang tool

For our first test project, we will create a simple Clang tool that runs the compiler and checks the syntax for the provided source file. We will create a so-called out-of-tree LLVM project, that is, a project that will use LLVM but will be located outside the main LLVM source tree.

Several actions are required to create the project:

- The required LLVM libraries and headers have to be built and installed

- We have to create a build configuration file for our test project

- The source code that uses LLVM has to be created

We will start with the first step and install the Clang support libraries and headers. We will use the following configuration command for CMake:

```
cmake -G Ninja -DCMAKE_BUILD_TYPE=Debug -DCMAKE_INSTALL_PREFIX=../install
  ↪  -DLLVM_TARGETS_TO_BUILD="X86" -DLLVM_ENABLE_PROJECTS="clang"
  ↪  -DLLVM_USE_LINKER=gold -DLLVM_USE_SPLIT_DWARF=ON -DBUILD_SHARED_LIBS=ON
  ↪  ../llvm
```

Figure 1.12: LLVM CMake configuration for a simple syntax checking Clang tool

As you may have noticed, we enabled only one project: clang. All other options are standard for our debug build. The command has to be run from a created build folder inside the LLVM source tree, as was suggested in *Section 1.3.1, Configuration with CMake*.

> **Important note**
>
> The configuration specified in Figure 1.12 will be the default build configuration
> used throughout the book.
>
> The configuration with shared libraries, in addition to the reduced size, has the
> advantage of simplifying the specification of dependencies. You only need to specify
> the shared libraries that your project directly depends on, and the dynamic linker
> takes care of the rest.

The required libraries and headers can be installed with the following command:

```
$ ninja install
```

The libraries and headers will be installed into install folder, as was specified by the
CMAKE_INSTALL_PREFIX option.

We have to create two files for our project:

- CMakeLists.txt: The project configuration file

- TestProject.cpp: The project source code

The project configuration file, CMakeLists.txt , will accept a path to the LLVM install
folder via the LLVM_HOME environment variable. The file is as follows:

```
1 cmake_minimum_required(VERSION 3.16)
2 project("syntax-check")
3
4 if ( NOT DEFINED ENV{LLVM_HOME})
5   message(FATAL_ERROR "$LLVM_HOME is not defined")
6 else()
7   message(STATUS "$LLVM_HOME found: $ENV{LLVM_HOME}")
8   set(LLVM_HOME $ENV{LLVM_HOME} CACHE PATH "Root of LLVM installation")
9   set(LLVM_LIB ${LLVM_HOME}/lib)
```

```
10    set(LLVM_DIR ${LLVM_LIB}/cmake/llvm)
11    find_package(LLVM REQUIRED CONFIG)
12    include_directories(${LLVM_INCLUDE_DIRS})
13    link_directories(${LLVM_LIBRARY_DIRS})
14    set(SOURCE_FILES SyntaxCheck.cpp)
15    add_executable(syntax-check ${SOURCE_FILES})
16    set_target_properties(syntax-check PROPERTIES COMPILE_FLAGS "-fno-rtti")
17    target_link_libraries(syntax-check
18      LLVMSupport
19      clangBasic
20      clangFrontend
21      clangSerialization
22      clangTooling
23    )
24 endif()
```

Figure 1.13: CMake file for simple syntax check Clang Tool

The most important parts of the file are as follows:

- *Line 2*: We specify the project name (syntax-check). That is also the name of our executable.

- *Lines 4-7*: Test for the `LLVM_HOME` environment variable.

- *Line 10*: We set a path to the LLVM CMake helpers.

- *Line 11*: We load the LLVM CMake package from the paths specified on *Line 10*.

- *Line 14*: We specify our source file that should be compiled.

- *Line 16*: We set up an additional flag for compilation: `-fno-rtti`. The flag is required as soon as LLVM is built without RTTI. This is done in an effort to reduce code and executable size [11].

- *Lines 18-22* We specify the required libraries to be linked to our program.

The source code for our tool is as follows:

```
1 #include "clang/Frontend/FrontendActions.h" // clang::SyntaxOnlyAction
2 #include "clang/Tooling/CommonOptionsParser.h"
3 #include "clang/Tooling/Tooling.h"
4 #include "llvm/Support/CommandLine.h" // llvm::cl::extrahelp
5
6 namespace {
7 llvm::cl::OptionCategory TestCategory("Test project");
8 llvm::cl::extrahelp
9     CommonHelp(clang::tooling::CommonOptionsParser::HelpMessage);
10 } // namespace
11
12 int main(int argc, const char **argv) {
13   llvm::Expected<clang::tooling::CommonOptionsParser> OptionsParser =
14       clang::tooling::CommonOptionsParser::create(argc, argv,
          ↳ TestCategory);
15   if (!OptionsParser) {
16     llvm::errs() << OptionsParser.takeError();
17     return 1;
18   }
19   clang::tooling::ClangTool Tool(OptionsParser->getCompilations(),
20                                  OptionsParser->getSourcePathList());
21   return Tool.run(
22       clang::tooling::newFrontendActionFactory<clang::SyntaxOnlyAction>()
23           .get());
24 }
```

Figure 1.14: SyntaxCheck.cpp

The most important part of the file are as follows:

- *Lines 7-9*: The majority of compiler tools have the same set of command line arguments. The LLVM command-line library [12] provides some APIs to process compiler command options. We set up the library on *Line 7*. We also set up additional help messages on lines 8-10.

- *Lines 13-18*: We parse command-line arguments.

- *Lines 19-24*: We create and run our Clang tool.

- *Lines 22-23*: We use the `clang::SyntaxOnlyAction` frontend action, which will run syntax and semantic checks on the input file. You can get more info about frontend actions in *Section 2.4.1, Frontend action.*

We have to specify a path to the LLVM `install` folder to build our tool. As was mentioned earlier, the path has to be specified via the `LLVM_HOME` environment variable. Our configuration command (see Figure 1.12) specifies the path as the `install` folder inside the LLVM project source tree. Thus we can build our tool as follows:

```
export LLVM_HOME=<...>/llvm-project/install
mkdir build
cd build
cmake -G Ninja ..
ninja
```

Figure 1.15: The syntax-check build commands

We can run the tool as follows:

```
$ cd build
$ ./syntax-check --help
USAGE: syntax-check [options] <source0> [... <sourceN>]

...
```

Figure 1.16: The syntax-check –help output

The program will successively terminate if we run it on a valid C++ source file, but it will produce an error message if it's run on a broken C++ file:

```
$ ./syntax-check mainbroken.cpp -- -std=c++17

mainbroken.cpp:2:11: error: expected ';' after return statement

  return 0

          ^

          ;

1 error generated.

Error while processing mainbroken.cpp.
```

Figure 1.17: The syntax-check run on a file with a syntax error

We used '- -' to pass additional arguments to the compiler in Figure 1.17, specifically indicating that we want to use C++17 with the option '-std=c++17'.

We can also run our tool with the LLDB debugger:

```
$   <...>/llvm-project/install/bin/lldb \

                ./syntax-check \

                --                 \

                main.cpp           \

                -- -std=c++17
```

Figure 1.18: The syntax-check run under debugger

We run syntax-check as the primary binary and set main.cpp source file as an argument for the tool (Figure 1.18). We also pass additional compilation flags (-std=c++17) to the syntax-check executable.

We can set a breakpoint and run the program as follows:

```
 1 (lldb) b clang::ParseAST
 2 ...
 3 (lldb) r
 4 ...
 5 Running without flags.
 6 Process 608249 stopped
 7 * thread #1, name = 'syntax-check', stop reason = breakpoint 1.1
 8     frame #0: ... clang::ParseAST(...) at ParseAST.cpp:117:3
 9     114
10     115  void clang::ParseAST(Sema &S, bool PrintStats, bool
   ↪  SkipFunctionBodies) {
11     116    // Collect global stats on Decls/Stmts (until we have a module
   ↪  streamer).
12 -> 117    if (PrintStats) {
13     118        Decl::EnableStatistics();
14     119        Stmt::EnableStatistics();
15     120    }
16 (lldb) c
17 Process 608249 resuming
18 Process 608249 exited with status = 0 (0x00000000)
19 (lldb)
```

Figure 1.19: LLDB session for Clang Tool test project

We set a breakpoint in the `clang::ParseAST` function (Figure 1.19, line 1). The function is the primary entry point for source code parsing. We run the program on *Line 3* and continue the execution after the breakpoint on *Line 16*.

We will use the same debugging techniques later in the book when we investigate Clang's source code.

1.5   Summary

In this chapter, we covered the history of the LLVM project, obtained the source code for LLVM, and explored its internal structure. We learned about the tools used to build LLVM, such as CMake and Ninja. We studied the various configuration options for building LLVM and how they can be used to optimize resources, including disk space. We built Clang and LLDB in debug and release modes and used the resulting tools to compile a basic program and run it with the debugger. We also created a simple Clang tool and ran it with the LLDB debugger.

The next chapter will introduce you to the compiler design architecture and explain how it appears in the context of Clang . We will primarily focus on the Clang frontend, but we will also cover the important concept of the Clang driver – the backbone that manages all stages of the compilation process, from parsing to linking.

1.6   Further reading

- Getting Started with the LLVM System: `https://llvm.org/docs/GettingStarted.html`

- Building LLVM with CMake: `https://llvm.org/docs/CMake.html`

- Clang Compiler User's Manual: `https://clang.llvm.org/docs/UsersManual.html`

2

Clang Architecture

In this chapter, we will examine the internal architecture of Clang and its relationship with other LLVM components. We will begin with an overview of the overall compiler architecture, with a specific focus on the clang driver. As the backbone of the compiler, the driver runs all compilation phases and controls their execution. Finally, we will concentrate on the frontend portion of the Clang compiler, which includes lexical and semantic analysis, and produces an **Abstract Syntax Tree (AST**) as its primary output. The AST forms the foundation for most Clang tools, and we will examine it more closely in the next chapters.

The following topics will be covered in this chapter:

- Compiler overview

- Clang driver overview, including an explanation of the compilation phases and their execution

- Clang frontend overview covering the preprocessing step, parsing, and semantic analysis

2.1   Technical requirements

The source code for this chapter is located in the `chapter2` folder of the book's GitHub repository: `https://github.com/PacktPublishing/Clang-Compiler-Frontend-Packt/tree/main/chapter2`.

2.2   Getting started with compilers

Despite the fact that compilers are used to translate programs from one form to another, they can also be considered large software systems that use various algorithms and data structures. The knowledge obtained by studying compilers can be used to design other scalable software systems as well. On the other hand, compilers are also a subject of active scientific research, and there are many unexplored areas and topics to investigate.

You can find some basic information about the internal structure of a compiler here. We will keep it as basic as possible so the information applies to any compiler, not just Clang. We will briefly cover all phases of compilation, which will help to understand Clang's position in the overall compiler architecture.

2.2.1   Exploring the compiler workflow

The primary function of a compiler is to convert a program written in a specific programming language (such as C/C++ or FORTRAN) into a format that can be executed on a target platform. This process involves the use of a compiler, which takes the source file and any compilation flags, and produces a build artifact, such as an executable or object file, as shown in Figure 2.1.

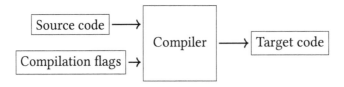

Figure 2.1: Compiler workflow

The term "target platform" can have a broad meaning. It can refer to machine code that is executed on the same host, as is typically the case. But it can also refer to cross-compilation, where the compiler generates code for a different computer architecture than the host. For example, code for a mobile application or embedded application running on ARM can be generated using an Intel machine as the host. Additionally, the target platform is not limited to machine code only. For example, some early C++ compilers (such as "cc") would produce pure C code as output. This was done because, at the time, C was the most widely used and well-established programming language, and the C compiler was the most reliable way to generate machine code. This approach allowed early C++ programs to be run on a wide range of platforms since most systems already had a C compiler available. The produced C code could then be compiled into machine code using any popular C Compiler such as GCC or LCC.

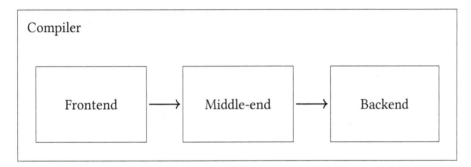

Figure 2.2: Typical compiler workflow: source program is passed via different stages: frontend, middle-end, and backend

We are going to focus on compilers that produce binary code, and a typical compiler workflow for such a compiler is shown in Figure 2.2. The stages of compilation can be described as follows:

- Frontend: The frontend does lexical analysis and parsing, which includes both syntax analysis and semantic analysis. The syntax analysis assumes that your program is well-organized according to the language grammar rules. The semantic analysis performs checks on the program's meaning and rejects invalid programs, such as those that use wrong types.

- Middle-end: The middle-end performs various optimizations on the intermediate representation (IR) code (LLVM-IR for Clang).

- Backend: The Backend of a compiler takes the optimized or transformed IR and generates machine code or assembly code that can be executed by the target platform.

The source program is transformed into different forms as it passes through the various stages. For example, the frontend produces IR code, which is then optimized by the middle-end, and finally converted into native code by the backend (see Figure 2.3).

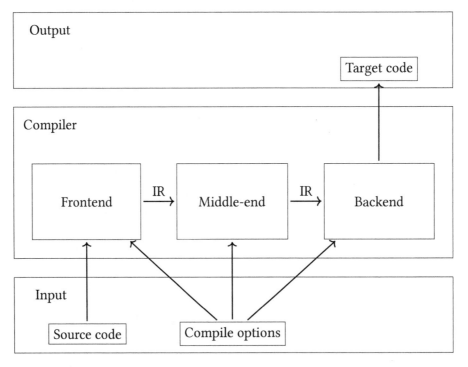

Figure 2.3: Source code transformation by compiler

Input data consists of **Source code** and **Compile options**. The source code is transformed by the **Frontend** into **IR**. The **Middle-end** does different optimizations on **IR** and passes the final (optimized) result to the **Backend**. The **Backend** generates the **Target code**. The **Frontend, Middle-end**, and **Backend** use **Compile options** as settings for the code transformations. Let's look into the compiler frontend as the first component of the compiler's workflow.

2.2.2 Frontend

The primary goal for the frontend is to convert a given source code to intermediate form. It's worth mentioning that the frontend also transforms the source code into various forms before it produces the IR. The frontend will be our primary focus in the book, so we will examine its components. The first component of the frontend is the Lexer (see Figure 2.4). It converts the source code into a set of tokens, which are used to create a special data structure called the abstract syntax tree (AST). The final component, the code generator (Codegen), traverses the AST and generates the IR from it.

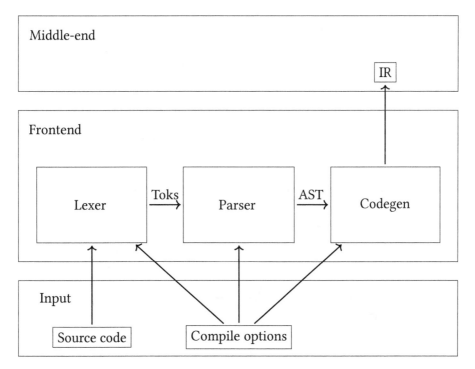

Figure 2.4: Compiler frontend

The source code is transformed into a set of tokens (**Toks**) by the **Lexer** . The **Parser** takes the tokens and creates an **Abstract Syntax Tree** (**AST**) that we will explore in details later in *Chapter 3, Clang AST.* The **Codegen** generates **IR** from the **AST** .

We will use a simple C/C++ program that calculates the maximum of two numbers to demonstrate the workings of the frontend. The code for the program is as follows:

```
1 int max(int a, int b) {
2   if (a > b)
3     return a;
4   return b;
5 }
```

Figure 2.5: Test program for compiler frontend investigations

The first component of the frontend is the lexer. Let's examine it.

Lexer

The frontend process starts with the Lexer , which converts the input source into a stream of tokens. In our example program (see Figure 2.5), the first token is the keyword int , which represents the integer type. This is followed by the identifier max for the function name. The next token is the left parenthesis (, and so on (see Figure 2.6).

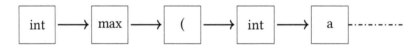

Figure 2.6: Lexer : the program source is converted into a stream of tokens

Parser

The Parser is the next component following the Lexer . The primary output produced by the Parser is called an **abstract syntax tree (AST)**. This tree represents the abstract syntactic structure of the source code written in a programming language. The Parser generates the AST by taking the stream of tokens produced by the Lexer as input and organizing them into a tree-like structure. Each node in the tree represents a construct in the source code, such as a statement or expression, and the edges between nodes represent the relationships between these constructs.

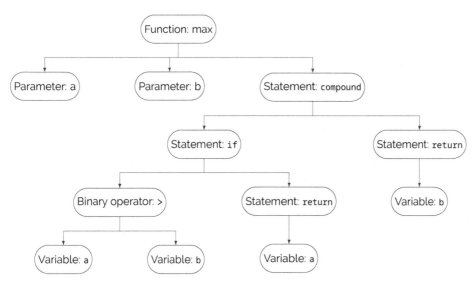

Figure 2.7: The AST for our example program, which calculates a maximum of two numbers

The AST for our example program is shown in Figure 2.7. As you can see, our function (max) has two parameters (a and b) and a body. The body is marked as a compound statement in Figure 2.7, see also Figure 2.40, where we provide a definition for a compound statement from the C++ standard. The compound statement consists of other statements, such as return and if . The a and b variables are used in the bodies of these statements. You may also be interested in the real AST generated by Clang for the compound statement, the result of which is shown in Figure 2.8.

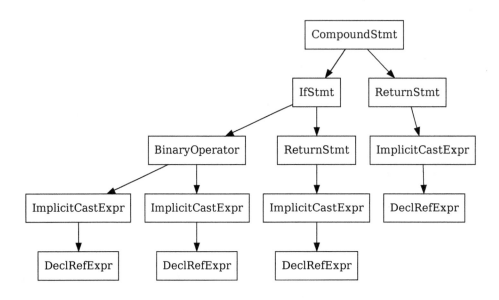

Figure 2.8: The AST for the compound statement generated by Clang . The tree generated by the clang -cc1 -ast-view <...> *command*

The Parser performs two activities:

1. Syntax analysis: the Parser constructs the AST by analyzing the syntax of the program.

2. Semantic analysis: the Parser analyzes the program semantically.

One of the jobs of the parser is to produce an error message if the parsing fails in either of the syntax or semantic analysis phases. If no error occurs, then we get a parse tree (or an AST) for the syntax analysis and a semantically verified parse tree in the case of semantic analysis. We can get a sense of this by considering what types of errors are detected by syntax analysis and which ones are detected by semantic analysis.

Syntax analysis assumes that the program should be correct in terms of the grammar specified for the language. For example, the following program is invalid in terms of syntax because a semicolon is missing from the last return statement:

```
1 int max(int a, int b) {
2   if (a > b)
3     return a;
4   return b // missing ;
5 }
```

Figure 2.9: Listing of program code with a syntax error

Clang produces the following output for the program:

```
max_invalid_syntax.cpp:4:11: error: expected ';' after return statement
  return b // missing ;
          ^
          ;
```

Figure 2.10: Compiler output for a program with a syntax error

On the other hand, a program can be syntactically correct but make no sense. The Parser should detect a semantic error in such cases. For instance, the following program has a semantic error related to the wrongly used type for the return value:

```
1 int max(int a, int b) {
2   if (a > b)
3     return a;
4   return &b; // invalid return type
5 }
```

Figure 2.11: Listing of program code with a semantic error

Clang generates the following output for the program:

```
max_invalid_sema.cpp:4:10: error: cannot initialize return object of type \
'int' with an rvalue of type 'int *'
  return &b; // invalid return type
         ^~
```

Figure 2.12: Compiler output for a program with a semantic error

AST is mainly constructed as a result of syntax analysis, but for certain languages, such as C++, semantic analysis is also crucial for constructing the AST, particularly for C++ template instantiation.

During syntax analysis, the compiler verifies that the template declaration adheres to the language's grammar and syntax rules, including the proper use of keywords such as "template" and "typename," as well as the formation of the template parameters and body.

Semantic analysis, on the other hand, involves the compiler performing template instantiation, which generates the AST for specific instances of the template. It's worth noting that the semantic analysis of templates can be quite complex, as the compiler must perform tasks such as type checking, name resolution, and more for each template instantiation. Additionally, the instantiation process can be recursive and lead to a significant amount of code duplication, known as code bloat. To combat this, C++ compilers employ techniques such as template instantiation caching to minimize the amount of redundant code generated.

The codegen

The codegen (it's worth mentioning that we also have another Codegen component as a part of Backend that generate the target code) or code generator, which is the final component of the compiler's frontend, has the primary goal of generating the **Intermediate Representation (IR)**. For this purpose, the compiler traverses the AST generated by the parser and converts it into other source code that is called the Intermediate Representation or IR. The IR is a language-independent representation, allowing the same middle-end component to be used for different frontends (FORTRAN vs C++). Another reason for using an Intermediate Representation (IR) is that if we have a new architecture available tomorrow, we can generate the target code specific to that architecture. Since the source language remains unchanged, all the steps leading up to the IR will remain the same. The IR provides this flexibility.

The use of IRs in compilers is a concept that has been around for several decades. The idea of using an intermediate representation to represent the source code of a program during compilation has evolved over time, and the exact date when IR was first introduced in compilers is not clear.

However, it is known that the first compilers in the 1950s and 1960s did not use IRs and instead translated source code directly into machine code. By the 1960s and 1970s, researchers had begun experimenting with using IRs in compilers to improve the efficiency and flexibility of the compilation process.

One of the first widely used IRs was three-address code, which was used in the mid-1960s in IBM/360's FORTRAN compiler. Other early examples of IRs include the **register transfer language (RTL)** and the **static single assignment (SSA)** form, which were introduced in the 1970s and 1980s respectively.

Today, the use of IRs in compilers is a standard practice, and many compilers use multiple IRs throughout the compilation process. This allows for more powerful optimization and code generation techniques to be applied.

2.3 Clang driver overview

When discussing compilers, we typically refer to a command-line utility that initiates and manages the compilation process. For example, to use the GNU Compiler Collection, one must call `gcc` to start the compilation process. Similarly, to compile a C++ program using Clang, one must call `clang` as the compiler. The program that controls the compilation process is known as the driver. The driver coordinates different stages of compilation and connects them together. In the book, we will be focusing on LLVM and using Clang as the driver for the compilation process.

It may be confusing for readers that the same word, "Clang," is used to refer to both the compiler frontend and the compilation driver. In contrast, with other compilers, where the driver and C++ compiler can be separate executables, "Clang" is a single executable that functions as both the driver and the compiler frontend. To use Clang as the compiler frontend only, the special option -cc1 must be passed to it.

2.3.1 Example program

We will use the simple "Hello world!" example program for our experiments with the Clang driver. The main source file is called hello.cpp . The file implements a trivial C++ program that prints "Hello world!" to the standard output:

```
1 #include <iostream>
2
3 int main() {
4   std::cout << "Hello world!" << std::endl;
5   return 0;
6 }
```

Figure 2.13: Example program: hello.cpp

You can compile the source with the following:

```
$ <...>/llvm-project/install/bin/clang hello.cpp -o /tmp/hello -lstdc++
```

Figure 2.14: Compilation for hello.cpp

As you can see, we used the clang executable as the compiler and specified the -lstdc++ library option because we used the <iostream> header from the standard C++ library. We also specified the output for the executable (/tmp/hello) with the -o option.

2.3.2 Compilation phases

We used two inputs for our example program. The first one is our source code and the second one is a shared library for the standard C++ library. The Clang driver should combine the inputs together, pass them via different phases of the compilation process, and finally, provide the executable file on the target platform.

Clang uses the same typical compiler workflow as shown in Figure 2.2. You can ask Clang to show the phases using the -ccc-print-phases additional argument:

```
$ <...>/llvm-project/install/bin/clang hello.cpp -o /tmp/hello -lstdc++ \
                                -ccc-print-phases
```

Figure 2.15: Command to print compilation phases for hello.cpp

The output for the command is the following:

```
            +- 0: input, "hello.cpp", c++
         +- 1: preprocessor, {0}, c++-cpp-output
       +- 2: compiler, {1}, ir
     +- 3: backend, {2}, assembler
  +- 4: assembler, {3}, object
|- 5: input, "1%dM", object
6: linker, {4, 5}, image
```

Figure 2.16: Compilation phases for hello.cpp

We can visualize the output as shown in Figure 2.17.

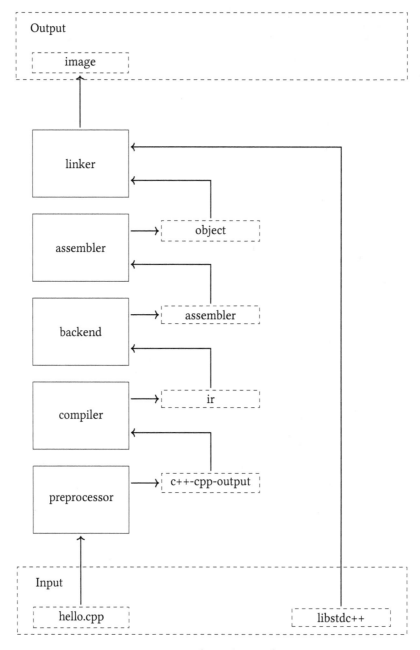

Figure 2.17: Clang driver phases

As we can see in Figure 2.17, the driver receives an input file, hello.cpp , which is a C++ file. The file is processed by the preprocessor and we obtain the preprocessor output (marked as c++-cpp-output). The result is compiled into IR form by the compiler, and then the backend converts it into assembly form. This form is later transformed into an object file. The final object file is combined with another object (libstdc++) to produce the final binary (image).

2.3.3 Tool execution

The phases are combined into several tool executions. The Clang driver invokes different programs to produce the final executable. Specifically, for our example, it calls the clang compiler and the ld linker. Both programs require additional arguments that are set up by the driver.

For instance, our example program (hello.cpp) includes the following header:

```
1 #include <iostream>
2 ...
```

Figure 2.18: iostream header at hello.cpp

We did not specify any additional arguments (such as search paths, for example, -I) when we invoked the compilation. However, different architectures and operating systems might have different paths for locating headers.

On Fedora 39, the header is located in the /usr/include/c++/13/iostream folder. We can examine a detailed description of the process executed by the driver and the arguments used with the -### option:

```
$ <...>/llvm-project/install/bin/clang hello.cpp -o /tmp/hello -lstdc++ -###
```

Figure 2.19: Command to print tools execution for hello.cpp

The output for this command is quite extensive, and certain parts have been omitted here. Please refer to Figure 2.20.

```
1  clang version 18.1.0rc (https://github.com/llvm/llvm-project.git ...)
2   "<...>/llvm-project/install/bin/clang-18"
3      "-cc1" ... \
4      "-internal-isystem" \
5      "/usr/include/c++/13" ... \
6      "-internal-isystem" \
7      "/usr/include/c++/13/x86_64-redhat-linux" ... \
8      "-internal-isystem" ... \
9      "<...>/llvm-project/install/lib/clang/18/include" ... \
10     "-internal-externc-isystem" \
11     "/usr/include" ... \
12     "-o" "/tmp/hello-XXX.o" "-x" "c++" "hello.cpp"
13  ".../bin/ld" ... \
14     "-o" "/tmp/hello" ... \
15     "/tmp/hello-XXX.o" \
16     "-lstdc++" ...
```

Figure 2.20: Clang driver tool execution. The host system is Fedora 39.

As we can see in Figure 2.20, the driver initiates two processes: clang-18 with the -cc1 flag (see *Lines 2-12*) and the linker ld (see *Lines 13-16*). The Clang compiler implicitly receives several search paths, as seen in *Lines 5, 7, 9, and 11*. These paths are necessary for the inclusion of the iostream header in the test program.

The output of the first executable (/tmp/hello-XXX.o) serves as input for the second one (see *Lines 12 and 15*). The arguments -lstdc++ and -o /tmp/hello are set for the linker, while the first argument (hello.cpp) is provided for the compiler invocation (first executable).

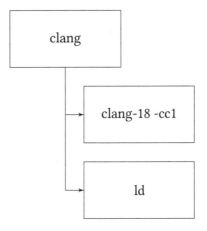

Figure 2.21: Clang driver tool execution. The Clang driver runs two executables: the clang executable with the -cc1 flag and the linker - ld executable

The process can be visualized as shown in Figure 2.21, where we can see that two executables are executed as part of the compilation process. The first one is clang-18 with a special flag (-cc1). The second one is the linker: ld .

2.3.4 Combining it all together

We can summarize the knowledge we have acquired so far using Figure 2.22. The figure illustrates two different processes started by the Clang driver. The first one is clang -cc1 (compiler), and the second one is ld (linker). The compiler process is the same executable as the Clang driver (clang), but it is run with a special argument: -cc1 . The compiler produces an object file that is then processed by the linker (ld) to generate the final binary.

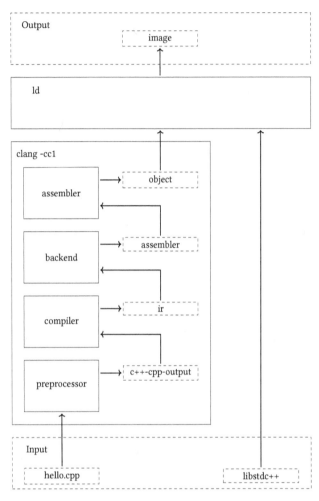

Figure 2.22: Clang driver: The driver got the input file `hello.cpp` *, which is a C++ file. It starts two processes:* `clang` *and* `ld` *. The first one does real compilation and starts the integrated assembler. The last one is the linker (* `ld` *) that produces the final binary (* `image` *) from the result received from the compiler and the external library (* `libstdc++` *)*

In Figure 2.22, we can observe similar components of the compiler mentioned earlier (see *Section 2.2, Getting started with compilers*). However, the main difference is that the **preprocessor** (part of the lexer) is shown separately, while the frontend and middle-end are combined into the **compiler**. Additionally, the figure depicts an **assembler** that is executed by the driver to generate the object code. It is important to note that the assembler can be integrated, as shown in Figure 2.22, or it may require a separate process to be executed.

> **Important note**
>
> Here is an example of specifying an external assembler using the -c (compile only) and -o (output file) options, along with the appropriate flags for your platform:
>
> ```
> $<...>/llvm-project/install/bin/clang -c hello.cpp \
>                                    -o /tmp/hello.o
> as -o /tmp/hello.o /tmp/hello.s
> ```

2.3.5 Debugging Clang

We're going to step through a debugging session for our compilation process, illustrated in Figure 2.14.

> **Important note**
>
> We will use the LLDB build created previously in *Section 1.3.3, The LLVM debugger, its build, and usage* for this and other debug sessions throughout the book. You can also use the LLDB that comes as part of your host system.

Our chosen point of interest, or breakpoint, is the clang::ParseAST function. In a typical debug session, which resembles the one outlined in Figure 1.11, you would feed command-line arguments following the "- -" symbol. The command should look like this:

```
$ lldb <...>/llvm-project/install/bin/clang -- hello.cpp -o /tmp/hello \
                                    -lstdc++
```

Figure 2.23: Running debugger for hello.cpp file compilation

In this case, <...> represents the directory path used to clone the LLVM project.

Unfortunately, this approach doesn't work with the Clang compiler:

```
1 $ lldb <...>/llvm-project/install/bin/clang -- hello.cpp -o /tmp/hello.o
  ↪ -lstdc++
2 ...
3 (lldb) b clang::ParseAST
4 ...
5 (lldb) r
6 ...
7 2 locations added to breakpoint 1
8 ...
9 Process 247135 stopped and restarted: thread 1 received signal: SIGCHLD
10 Process 247135 stopped and restarted: thread 1 received signal: SIGCHLD
11 Process 247135 exited with status = 0 (0x00000000)
12 (lldb)
```

Figure 2.24: Debugger session with failed interruption

As we can see from *Line 7*, the breakpoint was set but the process finished successfully (*Line 11*) without any interruptions. In other words, our breakpoint didn't trigger in this instance.

Understanding the internals of the Clang driver can help us identify the problem at hand. As mentioned earlier, the clang executable acts as a driver in this context, running two separate processes (refer to Figure 2.21). Therefore, if we wish to debug the compiler, we need to run it using the -cc1 option.

> **Important note**
>
> It's worth mentioning a certain optimization implemented in Clang in 2019 [22].
> When using the `-c` option, the Clang driver doesn't spawn a new process for the
> compiler:
>
> ```
> $ <...>/llvm-project/install/bin/clang -c hello.cpp    \
>                                        -o /tmp/hello.o \
>                                        -###
> clang version 18.1.0rc ...
> InstalledDir: <...>/llvm-project/install/bin
>  (in-process)
>  "<...>/llvm-project/install/bin/clang-18" "-cc1"..."hello.cpp"
>  ...
> ```
>
> As shown above, the Clang driver does not spawn a new process and instead calls
> the "cc1" tool within the same process. This feature not only improves the compiler's
> performance but can also be leveraged for Clang debugging.

Upon using the `-cc1` option and excluding the `-lstdc++` option (which is specific to the
second process, the ld linker), the debugger will generate the following output:

```
 1 $ lldb <...>/llvm-project/install/bin/clang -- -cc1 hello.cpp -o
   ↪ /tmp/hello.o
 2 ...
 3 (lldb) b clang::ParseAST
 4 ...
 5 (lldb) r
 6 ...
 7 2 locations added to breakpoint 1
 8 Process 249890 stopped
 9 * thread #1, name = 'clang', stop reason = breakpoint 1.1
10     frame #0: ... at ParseAST.cpp:117:3
11     114
12     115  void clang::ParseAST(Sema &S, bool PrintStats, bool
   ↪ SkipFunctionBodies) {
13     116    // Collect global stats on Decls/Stmts (until we have a module
   ↪ streamer).
14 -> 117    if (PrintStats) {
15     118      Decl::EnableStatistics();
16     119      Stmt::EnableStatistics();
17     120    }
18 (lldb) c
19 Process 249890 resuming
20 hello.cpp:1:10: fatal error: 'iostream' file not found
21 #include <iostream>
22          ^~~~~~~~~~
23 1 error generated.
24 Process 249890 exited with status = 1 (0x00000001)
25 (lldb)
```

Figure 2.25: Debugger session with missing search paths

Thus, we can see that we were able to successfully set the breakpoint, but the process ended with an error (see *Lines 20-24*). This error arose because we omitted certain search paths, which are typically appended implicitly by the Clang driver, necessary to find all the includes required for successful compilation.

We can successfully execute the process if we explicitly include all necessary arguments in the compiler invocation. Here's how to do that:

```
lldb <...>/llvm-project/install/bin/clang -- -cc1                       \
    -internal-isystem /usr/include/c++/13                               \
    -internal-isystem /usr/include/c++/13/x86_64-redhat-linux           \
    -internal-isystem <...>/llvm-project/install/lib/clang/18/include \
    -internal-externc-isystem /usr/include                             \
    hello.cpp                                                           \
    -o /tmp/hello.o
```

Figure 2.26: Running the debugger with specified search paths. Host system is Fedora 39

Then we can set the breakpoint for clang::ParseAST and run the debugger. The execution will complete without errors, as shown below:

```
 1 (lldb) b clang::ParseAST
 2 ...
 3 (lldb) r
 4 ...
 5 2 locations added to breakpoint 1
 6 Process 251736 stopped
 7 * thread #1, name = 'clang', stop reason = breakpoint 1.1
 8     frame #0: 0x00007fffe803eae0 ... at ParseAST.cpp:117:3
 9     114
10     115  void clang::ParseAST(Sema &S, bool PrintStats, bool
       ↪  SkipFunctionBodies) {
11     116    // Collect global stats on Decls/Stmts (until we have a module
       ↪  streamer).
12 -> 117    if (PrintStats) {
13     118      Decl::EnableStatistics();
14     119      Stmt::EnableStatistics();
15     120    }
16 (lldb) c
17 Process 251736 resuming
18 Process 251736 exited with status = 0 (0x00000000)
19 (lldb)
```

Figure 2.27: Successful debugger session for compiler

In conclusion, we have successfully demonstrated the debugging of a Clang compiler invocation. The techniques presented can be effectively employed for exploring the internals of a compiler and addressing compiler-related bugs.

2.4   Clang frontend overview

It's evident that the Clang compiler toolchain conforms to the pattern widely described in various compiler books [1, 18]. However, Clang's frontend part diverges significantly from a typical compiler frontend. The primary reason for this distinction is the complexity of the C++ language. Some features, such as macros, can modify the source code itself, while others, such as typedef, can influence the kind of token. Clang can also generate output in a variety of formats. For instance, the following command generates an aesthetically pleasing HTML view of the program shown in Figure 2.5:

```
$ <...>/llvm-project/install/bin/clang -cc1 -emit-html max.cpp
```

Take note that we pass the argument to emit the HTML form of the source program to the Clang frontend, specified with the `-cc1` option. Alternatively, you can pass an option to the frontend via the `-Xclang` option, which requires an additional argument representing the option itself, for example:

```
$ <...>/llvm-project/install/bin/clang -Xclang -emit-html max.cpp \
                            -fsyntax-only
```

You may notice that, in the preceding command, we utilized the `-fsyntax-only` option, instructing Clang to only execute the preprocessor, parser, and semantic analysis stages.

Accordingly, we can instruct the Clang frontend to perform different actions and produce varying types of output based on the provided compilation options. The base class for these actions is termed `FrontendAction` .

2.4.1 Frontend action

The Clang frontend is capable of executing only one frontend action at a time. A frontend action is a specific task or process that the frontend performs based on the provided compiler option. The following is a list of some possible frontend actions (the table only includes a subset of the available frontend actions):

FrontendAction	Compiler option	Description
EmitObjAction	`-emit-obj` (default)	Compile to an object file
EmitBCAction	`-emit-llvm-bc`	Compile to LLVM bytecode
EmitLLVMAction	`-emit-llvm`	Compile to LLVM readable form
ASTPrintAction	`-ast-print`	Build ASTs and then pretty-print them.
HTMLPrintAction	`-emit-html`	Prints the program source in HTML form
DumpTokensAction	`-dump-tokens`	Prints preprocessor tokens

Table 2.1: Frontend actions

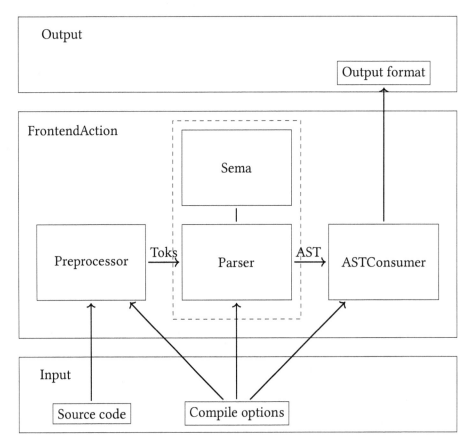

Figure 2.28: Clang frontend components

The diagram shown in Figure 2.28 illustrates the basic frontend architecture, which is similar to the architecture shown in Figure 2.4. However, there are notable differences specific to Clang.

One significant change is the naming of the lexer. In Clang, the lexer is referred to as the preprocessor. This naming convention reflects the fact that the lexer implementation is encapsulated within the `Preprocessor` class. This alteration was inspired by the unique aspects of the C/C++ language, which includes special types of tokens (macros) that require specialized preprocessing.

Another noteworthy deviation is found in the parser component. While conventional compilers typically perform both syntax and semantic analysis within the parser, Clang distributes these tasks across different components. The `Parser` component focuses solely on syntax analysis, while the `Sema` component handles semantic analysis.

Furthermore, Clang offers the ability to produce output in different forms or formats. For example, the `CodeGenAction` class serves as the base class for various code generation actions, such as `EmitObjAction` or `EmitLLVMAction`.

We will use the code for the max function from Figure 2.5 for our future exploration of the Clang frontend's internals:

```
1 int max(int a, int b) {
2   if (a > b)
3     return a;
4   return b;
5 }
```

Figure 2.29: Source code for max function: max.cpp

By utilizing the `-cc1` option, we can directly invoke the Clang frontend, bypassing the driver. This approach allows us to examine and analyze the inner workings of the Clang frontend in greater detail.

2.4.2 Preprocessor

The first part is the Lexer , which is called the preprocessor in Clang. Its primary goal is to convert the input program into a stream of tokens. You can print the token stream using the -dump-tokens options as follows:

```
$ <...>/llvm-project/install/bin/clang -cc1 -dump-tokens max.cpp
```

The output of the command is as shown:

```
int 'int'          [StartOfLine]  Loc=<max.cpp:1:1>
identifier 'max'             [LeadingSpace] Loc=<max.cpp:1:5>
l_paren '('                Loc=<max.cpp:1:8>
int 'int'                  Loc=<max.cpp:1:9>
identifier 'a'     [LeadingSpace] Loc=<max.cpp:1:13>
comma ','                  Loc=<max.cpp:1:14>
int 'int'          [LeadingSpace] Loc=<max.cpp:1:16>
identifier 'b'     [LeadingSpace] Loc=<max.cpp:1:20>
r_paren ')'                Loc=<max.cpp:1:21>
l_brace '{'        [LeadingSpace] Loc=<max.cpp:1:23>
if 'if'  [StartOfLine] [LeadingSpace]    Loc=<max.cpp:2:3>
l_paren '('        [LeadingSpace] Loc=<max.cpp:2:6>
identifier 'a'             Loc=<max.cpp:2:7>
greater '>'        [LeadingSpace] Loc=<max.cpp:2:9>
identifier 'b'     [LeadingSpace] Loc=<max.cpp:2:11>
r_paren ')'                Loc=<max.cpp:2:12>
return 'return'  [StartOfLine] [LeadingSpace]    Loc=<max.cpp:3:5>
identifier 'a'     [LeadingSpace] Loc=<max.cpp:3:12>
semi ';'                   Loc=<max.cpp:3:13>
return 'return'  [StartOfLine] [LeadingSpace]    Loc=<max.cpp:4:3>
identifier 'b'     [LeadingSpace] Loc=<max.cpp:4:10>
```

```
semi ';'                    Loc=<max.cpp:4:11>
r_brace '}'      [StartOfLine]  Loc=<max.cpp:5:1>
eof ''              Loc=<max.cpp:5:2>
```

Figure 2.30: Clang dump token output

As we can see, there are different types of tokens, such as language keywords (e.g., `int`, `return`), identifiers (e.g., max, a, b, etc.), and special symbols (e.g., semicolon, comma, etc.). The tokens for our small program are called **normal tokens**, which are returned by the lexer.

In addition to normal tokens, Clang has an additional type of token called **annotation tokens**. The primary difference is that these tokens also store additional semantic information. For instance, a sequence of normal tokens can be replaced by the parser with a single annotation token that contains information about the type or C++ scope. The primary reason for using such tokens is performance, as it allows for the prevention of reparsing when the parser needs to backtrack.

Since annotation tokens are used in the internal implementation of the parser, it would be good to consider an example of their usage with LLDB. Suppose we have the following C++ code:

```
1 namespace clangbook {
2 template <typename T> class A {};
3 } // namespace clangbook
4 clangbook::A<int> a;
```

Figure 2.31: Source code that uses annotation tokens, annotation.cpp

The last line of the code declares the variable a with the following type:

clangbook::A<int>. The type is represented as an annotation token, as shown in the following LLDB session:

```
 1  $ lldb <...>/llvm-project/install/bin/clang -- -cc1 annotation.cpp
 2  ...
 3  (lldb) b clang::Parser::ConsumeAnnotationToken
 4  ...
 5  (lldb) r
 6  ...
 7      608      }
 8      609
 9      610      SourceLocation ConsumeAnnotationToken() {
10  -> 611          assert(Tok.isAnnotation() && "wrong consume method");
11      612          SourceLocation Loc = Tok.getLocation();
12      613          PrevTokLocation = Tok.getAnnotationEndLoc();
13      614          PP.Lex(Tok);
14  (lldb) p Tok.getAnnotationRange().printToString(PP.getSourceManager())
15  (std::string) "<annotation.cpp:4:1, col:17>"
```

Figure 2.32: LLDB session for annotation.cpp

As we can see, Clang consumes an annotation token from *Line 4* of the program shown in Figure 2.31. The token is located between columns 1 and 7. See Figure 2.32. This corresponds to the following text used as the token: clangbook::A<int>. The token consists of other tokens, such as 'clangbook', '::', and so on. Combining all the tokens into one will significantly simplify the parsing and boost the overall parsing performance.

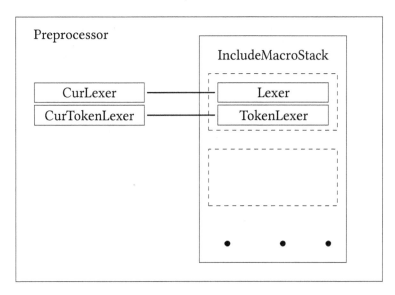

Figure 2.33: Preprocessor (clang lexer) class internals

C/C++ language has some specifics that influence the internal implementation of the `Preprocessor` class. The first one is about macros. The `Preprocessor` class has two different helper classes to retrieve tokens:

- The `Lexer` class is used to convert a text buffer into a stream of tokens.

- The `TokenLexer` class is used to retrieve tokens from macro expansions.

It should be noted that only one of these helpers can be active at a time.

Another specific aspect of C/C++ is the `#include` directive (which is also applicable to the import directive). In this case, we need to maintain a stack of includes, where each include can have its own `TokenLexer` or `Lexer`, depending on whether there is a macro expansion within it. As a result, the `Preprocessor` class keeps a stack of lexers (`IncludeMacroStack` class) for each `#include` directive, as shown in Figure 2.33.

2.4.3 Parser and sema

The parser and sema are crucial components of the Clang compiler frontend. They handle the syntax and semantic analysis of the source code, producing an AST as output. This tree can be visualized for our test program using the following command:

```
$ <...>/llvm-project/install/bin/clang -cc1 -ast-dump max.cpp
```

The output of this command is shown here:

```
TranslationUnitDecl 0xa9cb38 <<invalid sloc>> <invalid sloc>
|-TypedefDecl 0xa9d3a8 <<invalid sloc>> <invalid sloc>
implicit __int128_t '__int128'
| `-BuiltinType 0xa9d100 '__int128'
...
`-FunctionDecl 0xae6a98 <max.cpp:1:1, line:5:1> line:1:5 max
'int (int, int)'
  |-ParmVarDecl 0xae6930 <col:9, col:13> col:13 used a 'int'
  |-ParmVarDecl 0xae69b0 <col:16, col:20> col:20 used b 'int'
  `-CompoundStmt 0xae6cd8 <col:23, line:5:1>
    |-IfStmt 0xae6c70 <line:2:3, line:3:12>
    | |-BinaryOperator 0xae6c08 <line:2:7, col:11> 'bool' '>'
    | | |-ImplicitCastExpr 0xae6bd8 <col:7> 'int' <LValueToRValue>
    | | | `-DeclRefExpr 0xae6b98 <col:7> 'int' lvalue ParmVar 0xae6930
    | | |      'a' 'int'
    | | `-ImplicitCastExpr 0xae6bf0 <col:11> 'int' <LValueToRValue>
    | |   `-DeclRefExpr 0xae6bb8 <col:11> 'int' lvalue ParmVar 0xae69b0
    | |      'b' 'int'
    | `-ReturnStmt 0xae6c60 <line:3:5, col:12>
    |   `-ImplicitCastExpr 0xae6c48 <col:12> 'int' <LValueToRValue>
    |     `-DeclRefExpr 0xae6c28 <col:12> 'int' lvalue ParmVar 0xae6930
    |        'a' 'int'
    `-ReturnStmt 0xae6cc8 <line:4:3, col:10>
      `-ImplicitCastExpr 0xae6cb0 <col:10> 'int' <LValueToRValue>
```

```
`-DeclRefExpr 0xae6c90 <col:10> 'int' lvalue ParmVar 0xae69b0
    'b' 'int'
```

Figure 2.34: Clang AST dump output

Clang utilizes a hand-written recursive-descent parser [10]. This parser can be considered simple, and this simplicity was one key reason for its selection. Additionally, the complex rules specified for the C/C++ languages necessitated an ad hoc parser with easily adaptable rules.

Let's explore how this works with our example. Parsing begins with a top-level declaration known as a `TranslationUnitDecl`, representing a single translation unit. The C++ standard defines a translation unit as follows [21, lex.separate]:

> *A source file together with all the headers (16.5.1.2) and source files included (15.3) via the preprocessing directive #include, less any source lines skipped by any of the conditional inclusion (15.2) preprocessing directives, is called a translation unit.*

The parser first recognizes that the initial tokens from the source code correspond to a function definition, as defined in the C++ standard [21, dcl.fct.def.general]:

```
function-definition :
    ... declarator ... function-body
    ...
```

Figure 2.35: Function definition for C++ standard

The corresponding code follows:

```
1 int max(...) {
2   ...
3 }
```

Figure 2.36: Part of the example code corresponding to function definition from C++ standard

The function definition necessitates a declarator and function body. We'll start with the declarator, defined in the C++ standard as [21, dcl.decl.general]:

```
declarator:

        ...

        ... parameters-and-qualifiers ...

...

parameters-and-qualifiers:

        ( parameter-declaration-clause ) ...

...

parameter-declaration-clause:

        parameter-declaration-list ...

parameter-declaration-list:

        parameter-declaration

        parameter-declaration-list , parameter-declaration
```

Figure 2.37: Declarator definition for C++ standard

In other words, the declarator specifies a list of parameter declarations within brackets. The corresponding piece of code from the source is as follows:

```
1 ... (int a, int b)
2    ...
```

Figure 2.38: Part of the example code corresponding to declarator from C++ standard

The function definition, as stated above, also requires a function body. The C++ standard specifies the function body as follows: [21, dcl.fct.def.general]

```
function-body:

        ... compound-statement

        ...
```

Figure 2.39: Function body definition for C++ standard

Thus the function body consists of a compound statement, which is defined as follows in the C++ standard [21, stmt.block]:

```
compound-statement:
        { statement-seq ... }
statement-seq:
        statement
        statement-seq statement
```

Figure 2.40: Compound statement definition for C++ standard

Therefore, it describes a sequence of statements enclosed within {...} brackets.

Our program has two types of statements: the conditional (if) statement and the return statement. These are represented in the C++ grammar definition as follows [21, stmt.pre]:

```
statement:
        ...
        selection-statement
        ...
        jump-statement
        ...
```

Figure 2.41: Statement definition for C++ standard

In this context, the selection statement corresponds to the **if** condition in our program, while the jump statement corresponds to the **return** operator.

Let's examine the jum statement in more detail [21, stmt.jump.general]:

```
jump-statement:
        ...
        return expr-or-braced-init-list;
        ...
```

Figure 2.42: jump statement definition for C++ standard

where `expr-or-braced-init-list` is defined as [21, dcl.init.general]:

`expr-or-braced-init-list`:

      `expression`

       . . .

Figure 2.43: Return expression definition for C++ standard

In this context, the `return` keyword is followed by an expression and a semicolon. In our case, there's an implicit cast expression that automatically converts the variable into the required type (`int`).

It can be enlightening to examine the parser's operation through the LLDB debugger:

```
$ lldb <...>/llvm-project/install/bin/clang -- -cc1 max.cpp
```

The debugger session output is shown in Figure 2.44. As you can see, on *Line 1*, we've set a breakpoint for the parsing of return statements. Our program has two return statements. We bypass the first call (line 4) and halt at the second method invocation (*Line 9*). The backtrace (from the 'bt' command at *Line 13*) displays the call stack for the parsing process. This stack mirrors the parsing blocks we described earlier, adhering to the C++ grammar detailed in [21, lex.separate].

```
 1 (lldb) b clang::Parser::ParseReturnStatement
 2 (lldb) r
 3 ...
 4 (lldb) c
 5 ...
 6 * thread #1, name = 'clang', stop reason = breakpoint 1.1
 7     frame #0: ... clang::Parser::ParseReturnStatement(...) ...
 8   2421 StmtResult Parser::ParseReturnStatement() {
 9 -> 2422   assert((Tok.is(tok::kw_return) || Tok.is(tok::kw_co_return)) &&
10   2423          "Not a return stmt!");
11   2424   bool IsCoreturn = Tok.is(tok::kw_co_return);
12   2425   SourceLocation ReturnLoc = ConsumeToken();  // eat the 'return'.
13 (lldb) bt
14   * frame #0: ... clang::Parser::ParseReturnStatement( ...
15     ...
16     frame #2: ... clang::Parser::ParseStatementOrDeclaration( ...
17     frame #3: ... clang::Parser::ParseCompoundStatementBody( ...
18     frame #4: ... clang::Parser::ParseFunctionStatementBody( ...
19     frame #5: ... clang::Parser::ParseFunctionDefinition( ...
20     ...
```

Figure 2.44: Second return statement parsing at max.cpp example program

The parsing results in the generation of AST. We can also inspect the process of AST creation using the debugger. To do this, we need to set a corresponding breakpoint at the `clang::ReturnStmt::Create` method:

```
 1 $ lldb <...>/llvm-project/install/bin/clang -- -cc1 max.cpp
 2 ...
 3 (lldb) b clang::ReturnStmt::Create
 4 (lldb) r
 5 ...
 6 (lldb) c
 7 ...
 8 * thread #1, name = 'clang', stop reason = breakpoint 1.1
 9     frame #0: ... clang::ReturnStmt::Create(...) at Stmt.cpp:1205:8
10    1202
11    1203 ReturnStmt *ReturnStmt::Create(const ASTContext &Ctx,
         ↪ SourceLocation RL,
12    1204                                  Expr *E, const VarDecl
         ↪ *NRVOCandidate) {
13 -> 1205   bool HasNRVOCandidate = NRVOCandidate != nullptr;
14    1206   ...
15    1207   ...
16    1208   return new (Mem) ReturnStmt(RL, E, NRVOCandidate);
17 (lldb) bt
18 * thread #1, name = 'clang', stop reason = breakpoint 1.1
19   * frame #0: ... clang::ReturnStmt::Create( ...
20     frame #1: ... clang::Sema::BuildReturnStmt( ...
21     frame #2: ... clang::Sema::ActOnReturnStmt( ...
22     frame #3: ... clang::Parser::ParseReturnStatement( ...
23     frame #4: ...
         ↪ clang::Parser::ParseStatementOrDeclarationAfterAttributes( ...
24     ...
```

Figure 2.45: Breakpoint at clang::ReturnStmt::Create

As can be seen, the AST node for the return statement is created by the Sema component.

The beginning of the return statement parser can be located in frame 4:

```
 1 (lldb) f 4
 2 frame #4: ... clang::Parser::ParseStatementOrDeclarationAfterAttributes(
     ↳  ...
 3    323        SemiError = "break";
 4    324        break;
 5    325     case tok::kw_return:              // C99 6.8.6.4:
       ↳  return-statement
 6 -> 326        Res = ParseReturnStatement();
 7    327        SemiError = "return";
 8    328        break;
 9    329     case tok::kw_co_return:           // C++ Coroutines: ...
10 (lldb)
```

Figure 2.46: Return statement parsing at debugger

As we can observe, there is a reference to the C99 standard [25] for the corresponding statement. The standard [25] provides a detailed description of the statement and the process for handling it.

The code assumes that the current token is of type tok::kw_return, and in this case, the parser invokes the relevant clang::Parser::ParseReturnStatement method.

While the process of AST node creation can vary across different C++ constructs, it generally follows the pattern displayed in Figure 2.47.

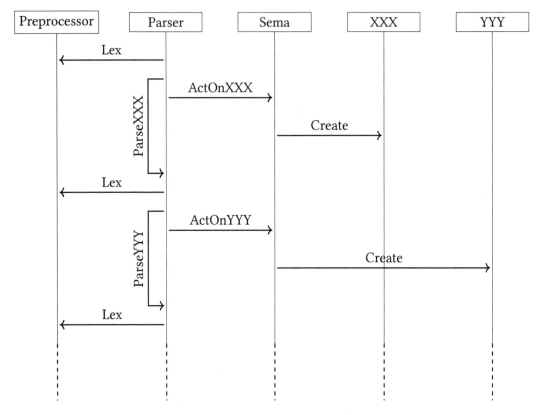

Figure 2.47: C++ parsing in Clang frontend

In Figure 2.47, the square boxes represent the corresponding classes, and the function calls are represented as edges with the called function shown as the edge label. As can be seen, the Parser invokes the Preprocessor::Lex method to retrieve a token from the lexer. It then calls a method corresponding to the token, for example, Parser:ParseXXX for the token XXX. This method then calls Sema::ActOnXXX, which creates the corresponding object using XXX::Create. The process is then repeated with a new token.

With this, we have now fully explored how the typical compiler frontend flow is implemented in Clang. We can see how the lexer component (the preprocessor) works in tandem with the parser (which comprises the parser and sema components) to produce the primary data structure for future code generation: the Abstract Syntax Tree (AST). The AST is not only essential for code generation but also for code analysis and modification. Clang provides easy access to the AST, thereby enabling the development of a diverse range of compiler tools.

2.5 Summary

In this chapter, we have acquired a basic understanding of compiler architecture and delved into the various stages of the compilation process, with a focus on the Clang driver. We have explored the internals of the Clang frontend, studying the Preprocessor that transforms a program into a set of tokens, and the Parser, which interacts with a component called 'Sema'. Together, these elements perform syntax and semantic analysis.

The upcoming chapter will center on the Clang Abstract Syntax Tree (AST)—the primary data structure employed in various Clang tools. We will discuss its construction and the methods for traversing it.

2.6 Further reading

- Working Draft, Standard for Programming Language C++: `https://eel.is/c++draft/`

- "Clang" CFE Internals Manual: `https://clang.llvm.org/docs/InternalsManual.html`

- Keith Cooper and Linda Torczon: Engineering A Compiler, 2012 [18]

3

Clang AST

The parsing stage of any compiler generates a parse tree, and the **Abstract Syntax Tree (AST)** is a fundamental algorithmic structure that is generated during the parsing of a given input program. The AST serves as the framework for the Clang frontend and is the primary tool for various Clang utilities, including linters. Clang offers sophisticated tools for searching (or matching) various AST nodes. These tools are implemented using a **Domain-Specific Language (DSL)**. It's crucial to understand its implementation to use it effectively.

We will start with the basic data structures and the class hierarchy that Clang uses to construct the AST. Additionally, we will explore the methods used for AST traversal and highlight some helper classes that facilitate node matching during this traversal. We will cover the following topics:

- Basic blocks used to construct the AST

- How the AST can be traversed

- The recursive visitor as the fundamental AST traversal tool

- AST matchers and their role in assisting with AST traversal

- Clang-Query as the basic tool to explore AST internals

- Compilation errors and their impact on the AST

3.1 Technical requirements

The source code for this chapter is located in the `chapter3` folder of the book's GitHub repository: `https://github.com/PacktPublishing/Clang-Compiler-Frontend-Packt /tree/main/chapter3`.

3.2 AST

The AST is usually depicted as a tree, with its leaf nodes corresponding to various objects, such as function declarations and loop bodies. Typically, the AST represents the result of syntax analysis, i.e., parsing. Clang's AST nodes were designed to be immutable. This design requires that the Clang AST stores results from semantic analysis, meaning the Clang AST represents the outcomes of both syntax and semantic analyses.

> **Important note**
>
> Although Clang also employs an AST, it's worth noting that the Clang AST is not a true tree. The presence of backward edges makes "graph" a more appropriate term for describing Clang's AST.

Typical tree structure implemented in C++ has all nodes derived from a base class. Clang uses a different approach. It splits different C++ constructions into separate groups with basic classes for each of them:

- Statements: `clang::Stmt` is the basic class for all statements. That includes ordinary statements such as `if` statements (`clang::IfStmt` class) as well as expressions and other C++ constructions.

- Declarations: `clang::Decl` is the base class for declarations. This includes a variable, typedef, function, struct, and more. There is also a separate base class for declarations with context, that is, declarations that might contain other declarations. The class is called clang::DeclContext. The declarations contained in clang::DeclContext can be accessed using the clang::DeclContext::decls method. Translation units (clang::TranslationUnitDecl class) and namespaces (clang::NamespaceDecl class) are typical examples of declarations with context.

- Types: C++ has a rich type system. It includes basic types such as int for integers as well as custom defined types and type redefinition via typedef or using. Types in C++ can have qualifiers such as const and can represent different memory addressing modes, aka pointers, references, and so on. Clang uses clang::Type as the basic class for type representations in AST.

It's worth noting that there are additional relations between the groups. For example, the clang::DeclStmt class, which inherits from clang::Stmt, has methods to retrieve corresponding declarations. Additionally, expressions (represented by the clang::Expr class), which inherit from clang::Stmt have methods to work with types. Let's look at all the groups in detail.

3.2.1 Statements

Stmt is the basic class for all statements. The statements can be combined into two sets (see Figure 3.1). The first one contains statements with values and the opposite group is for statements without values.

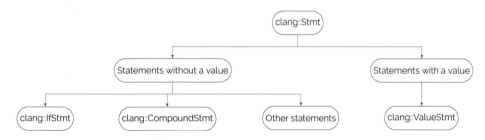

Figure 3.1: Clang AST: statements

The group of statements without a value consist of different C++ constructions such as if statements (`clang::IfStmt` class) or compound statements (`clang::CompoundStmt` class). The majority of all statements fall into the group.

The group of statements with a value consists of one base class `clang::ValueStmt` that has several children, such as `clang::LabelStmt` (for label representation) or `clang::ExprStmt` (for expression representation), see Figure 3.2.

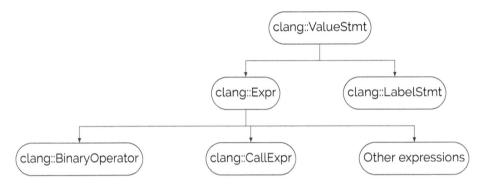

Figure 3.2: Clang AST: statements with a value

3.2.2 Declarations

Declarations can also be combined into two primary groups: declarations with context and without. Declarations with context can be considered as placeholders for other declarations. For example, a C++ namespace as well as a translation unit or function declaration might contain other declarations. A declaration of a friend entity (`clang::DeclFriend`) can be considered an example of a declaration without context.

It has to be noted that classes that are inherited from `DeclContext` also have `clang::Decl` as their top parent.

Some declarations can be redeclared, as in the following example:

```
1 extern int a;
2 int a = 1;
```

Figure 3.3: Declarations example: redeclaration.cpp

Such declarations have an additional parent that is implemented via a `clang::Redeclarable<...>` template.

3.2.3 Types

C++ is a statically typed language, which means that the types of variables must be declared at compile time. The types allow the compiler to make a reasonable conclusion about the program's meaning, which makes types an important part of semantic analysis. `clang::Type` is the basic class for types in Clang.

Types in C/C++ might have qualifiers that are called CV-qualifiers, as specified in the standard [21, basic.type.qualifier]. CV here stands for two keywords `const` and `volatile` that can be used as the qualifier for a type.

> **Important note**
>
> The C99 standard has an additional type qualifier, `restrict`, which is also supported by Clang [25, 6.7.3]. The type qualifier indicates to the compiler that, for the lifetime of the pointer, no other pointer will be used to access the object it points to. This allows the compiler to perform optimizations such as vectorization that wouldn't be possible otherwise. `restrict` helps limit pointer aliasing effects, which occur when multiple pointers reference the same memory location, thereby aiding optimizations. However, if the programmer's declaration of intent is not followed, and the object is accessed by an independent pointer, it results in undefined behavior.

Clang has a special class to support a type with a qualifier, `clang::QualType`, which is a pair of a pointer to `clang::Type` and a bit mask with information about the type qualifier. The class has a method to retrieve a pointer to the `clang::Type` and check different qualifiers. The following code (LLVM 18.x, `clang/lib/AST/ExprConstant.cpp`, *Line 3918*) shows how we can check a type for a const qualifier:

```cpp
bool checkConst(QualType QT) {
  // Assigning to a const object has undefined behavior.
  if (QT.isConstQualified()) {
    Info.FFDiag(E, diag::note_constexpr_modify_const_type) << QT;
    return false;
  }
  return true;
}
```

Figure 3.4: checkConst implementation from clang/lib/AST/ExprConstant.cpp

It's worth mentioning that clang::QualType has **operator->()** and **operator*()** implemented, that is, it can be considered as a smart pointer for the underlying clang::Type class.

In addition to qualifiers, the type can have additional information that represents different memory address models. For instance, there can be a pointer to an object or reference. clang::Type has the following helper methods to check different address models:

- clang::Type::isPointerType() for pointer type check

- clang::Type::isReferenceType() for reference type check

Types in C/C++ can also use aliases, which are introduced by using the **typedef** or **using** keywords. The following code defines foo and bar as aliases for the **int** type.

```cpp
1 using foo = int;
2 typedef int bar;
```

Figure 3.5: Type alias declarations

Original types, **int** in our case, are called canonical. You can test whether the type is canonical or not using the clang::QualType::isCanonical() method. clang::QualType also provides a method to retrieve the canonical type from an alias: `clang::QualType::getCanonicalType()`.

After gaining knowledge of the basic blocks used for the AST in Clang, it's time to investigate how these blocks can be used for AST traversal. This is the basic operation used by the compiler and compiler tools, and we will use it extensively throughout the book.

3.3   AST traversal

The compiler requires traversal of the AST to generate IR code. Thus, having a well-structured data structure for tree traversal is paramount for AST design. To put it another way, the design of the AST should prioritize facilitating easy tree traversal. A standard approach in many systems is to have a common base class for all AST nodes. This class typically provides a method to retrieve the node's children, allowing for tree traversal using popular algorithms such as Breadth-First Search (BFS) [19]. Clang, however, takes a different approach: its AST nodes don't share a common ancestor. This poses the question: how is tree traversal organized in Clang?

Clang employs three unique techniques:

- The Curiously Recurring Template Pattern (CRTP) for visitor class definition

- Ad hoc methods tailored specifically for different nodes

- Macros, which can be perceived as the connecting layer between the ad hoc methods and CRTP

We will explore these techniques through a simple program designed to identify function definitions and display the function names together with their parameters.

3.3.1   DeclVisitor test tool

Our test tool will build upon the `clang::DeclVisitor` class, which is defined as a straightforward visitor class aiding in the creation of visitors for C/C++ declarations.

We will use the same CMake file as was created for our first Clang tool (see Figure 1.13).

The sole addition to the new tool is the `clangAST` library. The resultant `CMakeLists.txt` is shown in Figure 3.6:

```
2  project("declvisitor")
3
4  if ( NOT DEFINED ENV{LLVM_HOME})
5    message(FATAL_ERROR "$LLVM_HOME is not defined")
6  else()
7    message(STATUS "$LLVM_HOME found: $ENV{LLVM_HOME}")
8    set(LLVM_HOME $ENV{LLVM_HOME} CACHE PATH "Root of LLVM installation")
9    set(LLVM_LIB ${LLVM_HOME}/lib)
10   set(LLVM_DIR ${LLVM_LIB}/cmake/llvm)
11   find_package(LLVM REQUIRED CONFIG)
12   include_directories(${LLVM_INCLUDE_DIRS})
13   link_directories(${LLVM_LIBRARY_DIRS})
14   set(SOURCE_FILE DeclVisitor.cpp)
15   add_executable(declvisitor ${SOURCE_FILE})
16   set_target_properties(declvisitor PROPERTIES COMPILE_FLAGS "-fno-rtti")
17   target_link_libraries(declvisitor
18     LLVMSupport
19     clangAST
20     clangBasic
21     clangFrontend
22     clangSerialization
23     clangTooling
24   )
```

Figure 3.6: CMakeLists.txt file for DeclVisitor test tool

The main function of our tool is presented below:

```
1 #include "clang/Tooling/CommonOptionsParser.h"
2 #include "clang/Tooling/Tooling.h"
3 #include "llvm/Support/CommandLine.h" // llvm::cl::extrahelp
4
5 #include "FrontendAction.hpp"
6
7 namespace {
8 llvm::cl::OptionCategory TestCategory("Test project");
9 llvm::cl::extrahelp
10     CommonHelp(clang::tooling::CommonOptionsParser::HelpMessage);
11 } // namespace
12
13 int main(int argc, const char **argv) {
14   llvm::Expected<clang::tooling::CommonOptionsParser> OptionsParser =
15       clang::tooling::CommonOptionsParser::create(argc, argv,
          ↪ TestCategory);
16   if (!OptionsParser) {
17     llvm::errs() << OptionsParser.takeError();
18     return 1;
19   }
20   clang::tooling::ClangTool Tool(OptionsParser->getCompilations(),
21                                  OptionsParser->getSourcePathList());
22   return Tool.run(clang::tooling::newFrontendActionFactory<
23                     clangbook::declvisitor::FrontendAction>()
24                   .get());
25 }
```

Figure 3.7: The main function of the DeclVisitor test tool

From *Lines 5 and 23*, it's evident that we employ a custom frontend action specific to our project: clangbook::declvisitor::FrontendAction.

The following is the code for this class:

```
1 #include "Consumer.hpp"
2 #include "clang/Frontend/FrontendActions.h"
3
4 namespace clangbook {
5 namespace declvisitor {
6 class FrontendAction : public clang::ASTFrontendAction {
7 public:
8   virtual std::unique_ptr<clang::ASTConsumer>
9   CreateASTConsumer(clang::CompilerInstance &CI,
10                     llvm::StringRef File) override {
11     return std::make_unique<Consumer>();
12   }
13 };
14 } // namespace declvisitor
15 } // namespace clangbook
```

Figure 3.8: Custom FrontendAction class for the DeclVisitor test tool

You'll notice that we have overridden the CreateASTConsumer function from clang::ASTFrontendAction class to instantiate an object of our custom AST consumer class Consumer, defined in clangbook::declvisitor namespace, as highlighted in Figure 3.8, *Lines 9-12.*

The implementation for the class is as follows:

```
1 #include "Visitor.hpp"
2 #include "clang/Frontend/ASTConsumers.h"
3
4 namespace clangbook {
5 namespace declvisitor {
6 class Consumer : public clang::ASTConsumer {
7 public:
8   Consumer() : V(std::make_unique<Visitor>()) {}
9
10   virtual void HandleTranslationUnit(clang::ASTContext &Context) override {
11     V->Visit(Context.getTranslationUnitDecl());
12   }
13
14 private:
15   std::unique_ptr<Visitor> V;
16 };
17 } // namespace declvisitor
18 } // namespace clangbook
```

Figure 3.9: Consumer class for the DeclVisitor test tool

Here, we can see that we've created a sample visitor and invoked it using an overridden method HandleTranslationUnit from the clang::ASTConsumer class (see Figure 3.9, *Line 11*).

However, the most intriguing portion is the code for the visitor:

```
1 #include "clang/AST/DeclVisitor.h"
2
3 namespace clangbook {
4 namespace declvisitor {
5 class Visitor : public clang::DeclVisitor<Visitor> {
6 public:
7   void VisitFunctionDecl(const clang::FunctionDecl *FD) {
8     llvm::outs() << "Function: '" << FD->getName() << "'\n";
9     for (auto Param : FD->parameters()) {
10       Visit(Param);
11     }
12   }
13   void VisitParmVarDecl(const clang::ParmVarDecl *PVD) {
14     llvm::outs() << "\tParameter: '" << PVD->getName() << "'\n";
15   }
16   void VisitTranslationUnitDecl(const clang::TranslationUnitDecl *TU) {
17     for (auto Decl : TU->decls()) {
18       Visit(Decl);
19     }
20   }
21 };
22 } // namespace declvisitor
23 } // namespace clangbook
```

Figure 3.10: Visitor class implementation

We will explore the code in more depth later. For now, we observe that it prints the function name at *Line 8* and the parameter name at *Line 14*.

We can compile our program using the same sequence of commands as we did for our test project, as detailed in *Section 1.4, Test project – syntax check with a Clang tool*.

```
export LLVM_HOME=<...>/llvm-project/install
mkdir build
cd build
cmake -G Ninja -DCMAKE_BUILD_TYPE=Debug ...
ninja
```

Figure 3.11: Configure and build commands for the DeclVisitor test tool

As you may notice, we used the -DCMAKE_BUILD_TYPE=Debug option for CMake. The option we are using will slow down the overall performance, but we use it because we might want to investigate the resulting program under debugger.

> **Important note**
>
> The build command we used for our tool assumes that the required libraries are installed under the <...>/llvm-project/install folder, which was specified with the -DCMAKE_INSTALL_PREFIX option during the CMake configure command, as described in *Section 1.4, Test project – syntax check with a Clang tool.* See Figure 1.12:
>
> ```
> cmake -G Ninja -DCMAKE_BUILD_TYPE=Debug
>   ↪  -DCMAKE_INSTALL_PREFIX=../install -DLLVM_TARGETS_TO_BUILD="X86"
>   ↪  -DLLVM_ENABLE_PROJECTS="clang" -DLLVM_USE_LINKER=gold
>   ↪  -DLLVM_USE_SPLIT_DWARF=ON -DBUILD_SHARED_LIBS=ON ../llvm
> ```
>
> The required build artifacts must be installed using the ninja install command.

We will use the same program we referenced in our previous investigations (see Figure 2.5) to also study AST traversal:

```
1 int max(int a, int b) {
2   if (a > b)
3     return a;
4   return b;
5 }
```

Figure 3.12: Test program max.cpp

This program consists of a single function, max, which takes two parameters, a and b, and returns the maximum of the two.

We can run our program as follows:

```
$ ./declvisitor max.cpp -- -std=c++17

...

Function: 'max'
        Parameter: 'a'
        Parameter: 'b'
```

Figure 3.13: The result of running the declvisitor utility on a test file

> **Important note**
>
> We used '- -' to pass additional arguments to the compiler in Figure 3.13, specifically indicating that we want to use C++17 with the option '-std=c++17'. We can also pass other compiler arguments. An alternative is to specify the compilation database path with the '-p' option, as shown below:
>
> ```
> $ ./declvisitor max.cpp -p <path>
> ```
>
> Here, <path> is the path to the folder containing the compilation database. You can find more information about the compilation database in *Chapter 9, Appendix 1:*

> *Compilation Database.*

Let's investigate the `Visitor` class implementation in detail.

3.3.2 Visitor implementation

Let's delve into the `Visitor` code (see Figure 3.10). Firstly, you'll notice an unusual construct where our class is derived from a base class parameterized by our own class:

```
5  class Visitor : public clang::DeclVisitor<Visitor> {
```

Figure 3.14: Visitor class declaration

This construct is known as the Curiously Recurring Template Pattern, or CRTP.

The Visitor class has several callbacks that are triggered when a corresponding AST node is encountered. The first callback targets the AST node representing a function declaration:

```
7   void VisitFunctionDecl(const clang::FunctionDecl *FD) {
8     llvm::outs() << "Function: '" << FD->getName() << "'\n";
9     for (auto Param : FD->parameters()) {
10      Visit(Param);
11    }
12  }
```

Figure 3.15: FunctionDecl callback

As shown in Figure 3.15, the function name is printed on *Line 8*. Our subsequent step involves printing the names of the parameters. To retrieve the function parameters, we can utilize the `parameters()` method from the `clang::FunctionDecl` class. This method was previously mentioned as an ad hoc approach for AST traversal. Each AST node provides its own methods to access child nodes. Since we have an AST node of a specific type (i.e., `clang::FunctionDecl*`) as an argument, we can employ these methods.

The function parameter is passed to the Visit(...) method of the base class clang::DeclVisitor<>, as shown in *Line 12* of Figure 3.15. This call is subsequently transformed into another callback, specifically for the clang::ParmVarDecl AST node:

```
13    void VisitParmVarDecl(const clang::ParmVarDecl *PVD) {
14      llvm::outs() << "\tParameter: '" << PVD->getName() << "'\n";
15    }
```

Figure 3.16: ParmVarDecl callback

You might be wondering how this conversion is achieved. The answer lies in a combination of the CRTP and C/C++ macros. To understand this, we need to dive into the Visit() method implementation of the clang::DeclVisitor<> class. This implementation heavily relies on C/C++ macros, so to get a glimpse of the actual code, we must expand these macros. This can be done using the -E compiler option. Let's make some modifications to CMakeLists.txt and introduce a new custom target.

```
25    add_custom_command(
26      OUTPUT ${SOURCE_FILE}.preprocessed
27      COMMAND ${CMAKE_CXX_COMPILER} -E -I ${LLVM_HOME}/include
          ↪  ${CMAKE_CURRENT_SOURCE_DIR}/${SOURCE_FILE} >
          ↪  ${SOURCE_FILE}.preprocessed
28      DEPENDS ${SOURCE_FILE}
29      COMMENT "Preprocessing ${SOURCE_FILE}"
30    )
31    add_custom_target(preprocess ALL DEPENDS ${SOURCE_FILE}.preprocessed)
```

Figure 3.17: Custom target to expand macros

We can run the target as follows:

```
$ ninja preprocess
```

The resulting file can be located in the build folder specified earlier, named
`DeclVisitor.cpp.preprocessed`. The build folder containig the file was specified by us
earlier when executing the cmake command (see Figure 3.11). Within this file, the generated
code for the `Visit()` method appears as follows:

```
1  RetTy Visit(typename Ptr<Decl>::type D) {
2    switch (D->getKind()) {
3      ...
4      case Decl::ParmVar: return
          ↪ static_cast<ImplClass*>(this)->VisitParmVarDecl(static_cast<typename
          ↪ Ptr<ParmVarDecl>::type>(D));
5      ...
6    }
7  }
```

Figure 3.18: Generated code for Visit() method

This code showcases the use of the CRTP in Clang. In this context, CRTP is employed to
redirect back to our `Visitor` class, which is referenced as `ImplClass`. CRTP allows the base
class to call a method from an inherited class. This pattern can serve as an alternative to
virtual functions and offers several advantages, the most notable being performance-related.
Specifically, the method call is resolved at compile time, eliminating the need for a vtable
lookup associated with virtual method calls.

The code was generated using C/C++ macros, as demonstrated here. This particular code
was sourced from the `clang/include/clang/AST/DeclVisitor.h` header:

```
34  #define DISPATCH(NAME, CLASS) \
35    return
         ↪ static_cast<ImplClass*>(this)->Visit##NAME(static_cast<PTR(CLASS)>(D))
```

Figure 3.19: DISPATCH macro definition from clang/include/clang/AST/DeclVisitor.h

NAME from Figure 3.19 is replaced with the node name; in our case, it's ParmVarDecl.

DeclVisitor is used to traverse C++ declarations. Clang also has StmtVisitor and TypeVisitor to traverse statements and types, respectively. These are built on the same principles as we considered in our example with the declaration visitor. However, these visitors come with several issues. They can only be used with specific groups of AST nodes. For instance, DeclVisitor can only be used with descendants of the Decl class. Another limitation is that we are required to implement recursion. For example, we set up recursion to traverse the function declaration in lines 9-11 (Figure 3.10). The same recursion was employed to traverse declarations within the translation unit (see Figure 3.10, *Lines 17-19*). This brings up another concern: it's possible to miss some parts of the recursion. For instance, our code will not function correctly if the max function declaration is specified inside a namespace. To address such scenarios, we would need to implement an additional visit method specifically for namespace declarations.

These challenges are addressed by the recursive visitor, which we will discuss shortly.

3.4   Recursive AST visitor

Recursive AST visitors address the limitations observed with specialized visitors. We will create the same program, which searches for and prints function declarations along with their parameters, but we'll use a recursive visitor this time.

The CMakeLists.txt for recursive visitor test tool will be used in a similar way as before. Only the project name (*Lines 2 and 15-17* in Figure 3.20) and source filename (*Line 14* in Figure 3.20 were changed:

```
 1 cmake_minimum_required(VERSION 3.16)
 2 project("recursivevisitor")
 3
 4 if ( NOT DEFINED ENV{LLVM_HOME})
 5   message(FATAL_ERROR "$LLVM_HOME is not defined")
 6 else()
 7   message(STATUS "$LLVM_HOME found: $ENV{LLVM_HOME}")
 8   set(LLVM_HOME $ENV{LLVM_HOME} CACHE PATH "Root of LLVM installation")
 9   set(LLVM_LIB ${LLVM_HOME}/lib)
10   set(LLVM_DIR ${LLVM_LIB}/cmake/llvm)
11   find_package(LLVM REQUIRED CONFIG)
12   include_directories(${LLVM_INCLUDE_DIRS})
13   link_directories(${LLVM_LIBRARY_DIRS})
14   set(SOURCE_FILE RecursiveVisitor.cpp)
15   add_executable(recursivevisitor ${SOURCE_FILE})
16   set_target_properties(recursivevisitor PROPERTIES COMPILE_FLAGS
      ↪  "-fno-rtti")
17   target_link_libraries(recursivevisitor
18     LLVMSupport
19     clangAST
20     clangBasic
21     clangFrontend
22     clangSerialization
23     clangTooling
24   )
25 endif()
```

Figure 3.20: CMakeLists.txt file for the RecursiveVisitor test tool

The main function for our tool is similar to the 'DeclVisitor' one defined in Figure 3.7.

```
 1 #include "clang/Tooling/CommonOptionsParser.h"
 2 #include "clang/Tooling/Tooling.h"
 3 #include "llvm/Support/CommandLine.h" // llvm::cl::extrahelp
 4
 5 #include "FrontendAction.hpp"
 6
 7 namespace {
 8 llvm::cl::OptionCategory TestCategory("Test project");
 9 llvm::cl::extrahelp
10     CommonHelp(clang::tooling::CommonOptionsParser::HelpMessage);
11 } // namespace
12
13 int main(int argc, const char **argv) {
14   llvm::Expected<clang::tooling::CommonOptionsParser> OptionsParser =
15       clang::tooling::CommonOptionsParser::create(argc, argv,
16         ↪ TestCategory);
16   if (!OptionsParser) {
17     llvm::errs() << OptionsParser.takeError();
18     return 1;
19   }
20   clang::tooling::ClangTool Tool(OptionsParser->getCompilations(),
21                                  OptionsParser->getSourcePathList());
22   return Tool.run(clang::tooling::newFrontendActionFactory<
23                       clangbook::recursivevisitor::FrontendAction>()
24                   .get());
25 }
```

Figure 3.21: The main function for the RecursiveVisitor test tool

As you can see, we changed only the namespace name for our custom frontend action at *Line 23*.

The code for the frontend action and consumer is the same as in Figure 3.8 and Figure 3.9, with the only difference being the namespace change from declvisitor to recursivevisitor. The most interesting part of the program is the Visitor class implementation.

```
1 #include "clang/AST/RecursiveASTVisitor.h"
2
3 namespace clangbook {
4 namespace recursivevisitor {
5 class Visitor : public clang::RecursiveASTVisitor<Visitor> {
6 public:
7   bool VisitFunctionDecl(const clang::FunctionDecl *FD) {
8     llvm::outs() << "Function: '" << FD->getName() << "'\n";
9     return true;
10  }
11  bool VisitParmVarDecl(const clang::ParmVarDecl *PVD) {
12    llvm::outs() << "\tParameter: '" << PVD->getName() << "'\n";
13    return true;
14  }
15 };
16 } // namespace recursivevisitor
17 } // namespace clangbook
```

Figure 3.22: Visitor class implementation

There are several changes compared to the code for 'DeclVisitor' (see Figure 3.10). The first is that recursion isn't implemented. We've only implemented the callbacks for nodes of interest to us. A reasonable question arises: how is the recursion controlled? The answer lies in another change: our callbacks now return a boolean result. The false value indicates that the recursion should stop, while true signals the visitor to continue the traversal.

The program can be compiled using the same sequence of commands as we used previously. See Figure 3.11.

We can run our program as follows, see Figure 3.23:

```
$ ./recursivevisitor max.cpp -- -std=c++17

...

Function: 'max'
        Parameter: 'a'
        Parameter: 'b'
```

Figure 3.23: The result of running the recursivevisitor utility on a test file

As we can see, it produces the same result as we obtained with the DeclVisitor implementation. The AST traversal techniques considered so far are not the only ways for AST traversal. Most of the tools that we will consider later will use a different approach based on AST matchers.

3.5   AST matchers

AST matchers [16] provide another approach for locating specific AST nodes. They can be particularly useful in linters when searching for improper pattern usage or in refactoring tools when identifying AST nodes for modification.

We will create a simple program to test AST matches. The program will identify a function definition with the name max. We will use a slightly modified CMakeLists.txt file from the previous examples to include the libraries required to support AST matches:

```
1  cmake_minimum_required(VERSION 3.16)
2  project("matchvisitor")
3
4  if ( NOT DEFINED ENV{LLVM_HOME})
5    message(FATAL_ERROR "$LLVM_HOME is not defined")
6  else()
7    message(STATUS "$LLVM_HOME found: $ENV{LLVM_HOME}")
8    set(LLVM_HOME $ENV{LLVM_HOME} CACHE PATH "Root of LLVM installation")
9    set(LLVM_LIB ${LLVM_HOME}/lib)
10   set(LLVM_DIR ${LLVM_LIB}/cmake/llvm)
11   find_package(LLVM REQUIRED CONFIG)
12   include_directories(${LLVM_INCLUDE_DIRS})
13   link_directories(${LLVM_LIBRARY_DIRS})
14   set(SOURCE_FILE MatchVisitor.cpp)
15   add_executable(matchvisitor ${SOURCE_FILE})
16   set_target_properties(matchvisitor PROPERTIES COMPILE_FLAGS "-fno-rtti")
17   target_link_libraries(matchvisitor
18     LLVMFrontendOpenMP
19     LLVMSupport
20     clangAST
21     clangASTMatchers
22     clangBasic
23     clangFrontend
24     clangSerialization
25     clangTooling
26   )
27 endif()
```

Figure 3.24: CMakeLists.txt for AST matchers test tool

There are two additional libraries added: `LLVMFrontendOpenMP` and `clangASTMatchers` (see *Lines 18 and 21* in Figure 3.24). The `main` function for our tool looks like this:

```
1  #include "clang/Tooling/CommonOptionsParser.h"
2  #include "clang/Tooling/Tooling.h"
3  #include "llvm/Support/CommandLine.h" // llvm::cl::extrahelp
4  #include "MatchCallback.hpp"
5
6  namespace {
7  llvm::cl::OptionCategory TestCategory("Test project");
8  llvm::cl::extrahelp
9      CommonHelp(clang::tooling::CommonOptionsParser::HelpMessage);
10 } // namespace
11
12 int main(int argc, const char **argv) {
13   llvm::Expected<clang::tooling::CommonOptionsParser> OptionsParser =
14       clang::tooling::CommonOptionsParser::create(argc, argv,
          ↪  TestCategory);
15   if (!OptionsParser) {
16     llvm::errs() << OptionsParser.takeError();
17     return 1;
18   }
19   clang::tooling::ClangTool Tool(OptionsParser->getCompilations(),
20                                  OptionsParser->getSourcePathList());
21   clangbook::matchvisitor::MatchCallback MC;
22   clang::ast_matchers::MatchFinder Finder;
23   Finder.addMatcher(clangbook::matchvisitor::M, &MC);
24   return Tool.run(clang::tooling::newFrontendActionFactory(&Finder).get());
25 }
```

Figure 3.25: The main function for AST matchers test tool

As you can observe (*Lines 21-23*), we employ the `MatchFinder` class and define a custom callback (included via the header in *Line 4*) that outlines the specific AST node we intend to match. The callback is implemented as follows:

```
1 #include "clang/ASTMatchers/ASTMatchFinder.h"
2 #include "clang/ASTMatchers/ASTMatchers.h"
3
4 namespace clangbook {
5 namespace matchvisitor {
6 using namespace clang::ast_matchers;
7 static const char *MatchID = "match-id";
8 clang::ast_matchers::DeclarationMatcher M =
9     functionDecl(decl().bind(MatchID), matchesName("max"));
10
11 class MatchCallback : public
    ↪  clang::ast_matchers::MatchFinder::MatchCallback {
12 public:
13   virtual void
14   run(const clang::ast_matchers::MatchFinder::MatchResult &Result) final {
15     if (const auto *FD =
        ↪  Result.Nodes.getNodeAs<clang::FunctionDecl>(MatchID)) {
16       const auto &SM = *Result.SourceManager;
17       const auto &Loc = FD->getLocation();
18       llvm::outs() << "Found 'max' function at " << SM.getFilename(Loc) <<
          ↪  ":"
19                   << SM.getSpellingLineNumber(Loc) << ":"
20                   << SM.getSpellingColumnNumber(Loc) << "\n";
21     }
22   }
23 };
24
25 } // namespace matchvisitor
26 } // namespace clangbook
```

Figure 3.26: The match callback for the AST matchers test tool

The most crucial section of the code is located at lines 7-9. Each matcher is identified by an ID, which, in our case, is 'match-id'. The matcher itself is defined in *Lines 8-9*:

```
8 clang::ast_matchers::DeclarationMatcher M =
9     functionDecl(decl().bind(MatchID), matchesName("max"));
```

This matcher seeks a function declaration that has a specific name, using `functionDecl()`, as seen in `matchesName()`. We utilized a specialized Domain-Specific Language (DSL) to specify the matcher. The DSL is implemented using C++ macros. We can also create our own matchers, as will be shown in *Section 7.3.3, Check implementation*. It's worth noting that the recursive AST visitor serves as the backbone for AST traversal inside the matcher's implementation.

The program can be compiled using the same sequence of commands as we used previously. See Figure 3.11.

We will utilize a slightly modified version of the example shown in Figure 2.5, with an additional function added:

```
1 int max(int a, int b) {
2   if (a > b) return a;
3   return b;
4 }
5
6 int min(int a, int b) {
7   if (a > b) return b;
8   return a;
9 }
```

Figure 3.27: Test program minmax.cpp for AST matchers

When we run our test tool on the example, we will obtain the following output:

```
./matchvisitor minmax.cpp -- -std=c++17

...

Found the 'max' function at minmax.cpp:1:5
```

Figure 3.28: The result of running the matchvisitor utility on a test file

As we can see, it has located only one function declaration with the name specified for the matcher.

The DSL for matchers is typically employed in custom Clang tools, such as clang-tidy (as discussed in *Chapter 5, Clang-Tidy Linter Framework*), but it can also be used as a standalone tool. A specialized program called `clang-query` enables the execution of different match queries, which can be used to search for specific AST nodes in analyzed C++ code. Let's see how the tool works.

3.6   Explore Clang AST with clang-query

AST matchers are incredibly useful, and there's a utility that facilitates checking various matchers and analyzing the AST of your source code. This utility is known as `clang-query` tool. You can build and install this utility using the following command:

```
$ ninja install-clang-query
```

Figure 3.29: The clang-query installation

You can run the tool as follows:

```
$ <...>/llvm-project/install/bin/clang-query minmax.cpp
```

Figure 3.30: Running clang-query on a test file

We can use the match command as follows:

```
clang-query> match functionDecl(decl().bind("match-id"), matchesName("max"))
Match #1:
minmax.cpp:1:1: note: "match-id" binds here
int max(int a, int b) {
^~~~~~~~~~~~~~~~~~~~~~~
minmax.cpp:1:1: note: "root" binds here
int max(int a, int b) {
^~~~~~~~~~~~~~~~~~~~~~~
1 match.
clang-query>
```

Figure 3.31: Working with clang-query

Figure 3.31 demonstrates the default output, referred to as ′diag′ . Among several potential outputs, the most relevant one for us is ′dump′ . When the output is set to ′dump′ , clang-query will display the located AST node. For example, the following demonstrates how to match a function parameter named a :

```
clang-query> set output dump
clang-query> match parmVarDecl(hasName("a"))
Match #1:
Binding for "root":
ParmVarDecl 0x6775e48 <minmax.cpp:1:9, col:13> col:13 used a 'int'
Match #2:
Binding for "root":
ParmVarDecl 0x6776218 <minmax.cpp:6:9, col:13> col:13 used a 'int'
2 matches.
clang-query>
```

Figure 3.32: Working with clang-query using dump output

This tool proves useful when you wish to test a particular matcher or investigate a portion of the AST tree. We will utilize this tool to explore how Clang handles compilation errors.

3.7   Processing AST in the case of errors

One of the most interesting aspects of Clang pertains to error processing. Error processing encompasses error detection, the display of corresponding error messages, and potential error recovery. The latter is particularly intriguing in terms of the Clang AST. Error recovery occurs when Clang doesn't halt upon encountering a compilation error but continues to compile in order to detect additional issues.

Such behavior is beneficial for various reasons. The most evident one is user convenience. When programmers compile a program, they typically prefer to be informed about as many errors as possible in a single compilation run. If the compiler were to stop at the first error, the programmer would have to correct that error, recompile, then address the subsequent error, and recompile again, and so forth. This iterative process can be tedious and frustrating, especially with larger code bases or intricate errors. While this behavior is particularly useful for compiled languages such as C/C++, it's worth noting that interpreted languages also exhibit this behavior, which can assist users in handling errors step by step.

Another compelling reason centers on IDE integration, which will be discussed in more detail in *Chapter 8, IDE Support and Clangd*. IDEs offer navigation support coupled with an integrated compiler. We will explore clangd as one such tool. Editing code in IDEs often leads to compilation errors. Most errors are confined to specific sections of the code, and it might be suboptimal to cease navigation in such cases.

Clang employs various techniques for error recovery. For the syntax stage of parsing, it utilizes heuristics; for instance, if a user forgets to insert a semicolon, Clang may attempt to add it as part of the recovery process. The Recovery Phase can be abbreviated as DIRT where D stands for Delete a character (for example, an extra semicolon), I stands for Insert a character (as in the example presented), R stands for Replace (which replaces a character to match a particular token), and T stands for Transpose (rearranging two characters to match a token).

Clang performs full recovery if it's possible and produces an AST that corresponds to the modified file with all compilation errors fixed. The most interesting case is when full recovery is not possible, and Clang implements unique techniques to manage recovery while AST is created.

Consider a program (maxerr.cpp) that is syntactically correct but has a semantic error. For example, it might use an undeclared variable. In this program, refer to *Line 3* where the undeclared variable ab is used:

```
1 int max(int a, int b) {
2   if (a > b) {
3     return ab;
4   }
5   return b;
6 }
```

Figure 3.33: The maxerr.cpp test program with a semantic error – undeclared variable

We are interested in the AST result produced by Clang, and we will use `clang-query` to examine it, which can be run as follows:

```
$ <...>/llvm-project/install/bin/clang-query maxerr.cpp
...
maxerr.cpp:3:12: error: use of undeclared identifier 'ab'
    return ab;
           ^
```

Figure 3.34: Compilation error example

From the output, we can see that clang-query displayed a compilation error detected by the compiler. It's worth noting that, despite this, an AST was produced for the program, and we can examine it. We are particularly interested in the return statements and can use the corresponding matcher to highlight the relevant parts of the AST.

We will also set up the output to produce the AST and search for return statements that are of interest to us:

```
clang-query> set output dump
clang-query> match returnStmt()
```

Figure 3.35: Setting the matcher for return statement

The resulting output identifies two return statements in our program: the first match on *Line 5* and the second match on *Line 3*:

```
Match #1:
Binding for "root":
ReturnStmt 0x6b63230 <maxerr.cpp:5:3, col:10>
`-ImplicitCastExpr 0x6b63218 <col:10> 'int' <LValueToRValue>
  `-DeclRefExpr 0x6b631f8 <col:10> 'int' lvalue ParmVar 0x6b62ec8 'b' 'int'



Match #2:
Binding for "root":
ReturnStmt 0x6b631b0 <maxerr.cpp:3:5, col:12>
`-RecoveryExpr 0x6b63190 <col:12> '<dependent type>' contains-errors lvalue

2 matches.
```

Figure 3.36: ReturnStmt node matches at maxerr.cpp test program

As we can see, the first match corresponds to semantically correct code on *Line 5* and contains a reference to the a parameter. The second match is for *Line 3*, which has a compilation error. Notably, Clang has inserted a special type of AST node: RecoveryExpr. It's worth noting that, in certain situations, Clang might produce an incomplete AST. This can cause issues with Clang tools, such as lint checks. In instances of compilation errors, lint checks might yield unexpected results because Clang couldn't recover accurately from the compilation errors. We will revisit the problem when exploring the clang-tidy lint check framework in *Chapter 5, Clang-Tidy Linter Framework*.

3.8 Summary

We explored the Clang AST, a major instrument for creating various Clang tools. We learned about the architectural design principles chosen for the implementation of the Clang AST and investigated different methods for AST traversal. We delved into specialized traversal techniques, such as those for C/C++ declarations, and also looked into more universal techniques that employ recursive visitors and Clang AST matchers. Our exploration concluded with the `clang-query` tool and how it can be used for Clang AST exploration. Specifically, we used it to understand how Clang processes compilation errors.

The next chapter will discuss the basic libraries used in Clang and LLVM development. We will explore the LLVM code style and foundational Clang/LLVM classes, such as `SourceManager` and `SourceLocation`. We will also cover the TableGen library, which is used for code generation, and the LLVM Integration Test (LIT) framework.

3.9 Further reading

- How to write RecursiveASTVisitor: `https://clang.llvm.org/docs/RAVFrontendAction.html`

- AST Matcher Reference: `https://clang.llvm.org/docs/LibASTMatchersReference.html`

4

Basic Libraries and Tools

LLVM is written in the C++ language and, as of July 2022, it uses the C++17 version of the C++ standard [6]. LLVM actively utilizes functionality provided by the **Standard Template Library (STL)**. On the other hand, LLVM contains numerous internal implementations [13] for fundamental containers, primarily aimed at optimizing performance. For example, `llvm::SmallVector` has an interface similar to `std::vector` but features an internally optimized implementation. Hence, familiarity with these extensions is essential for anyone wishing to work with LLVM and Clang.

Additionally, LLVM has introduced other development tools such as **TableGen**, a **domain specific language (DSL)** designed for structural data processing, and **LIT** (LLVM Integrated Tester), the LLVM test framework. More details about these tools are discussed later in this chapter. We'll cover the following topics in this chapter:

- LLVM coding style

- LLVM basic libraries

- Clang basic libraries

- LLVM supporting tools

- Clang plugin project

We plan to use a simple example project to demonstrate these tools. This project will be a Clang plugin that estimates the complexity of C++ classes. A class is considered complex if the number of methods exceeds a threshold specified as a parameter. While this definition of complexity may be considered trivial, we will explore more advanced definitions of complexity later in *Chapter 6, Advanced Code Analysis*.

4.1   Technical requirements

The source code for this chapter is located in the `chapter4` folder of the book's GitHub repository: `https://github.com/PacktPublishing/Clang-Compiler-Frontend-Packt /tree/main/chapter4`.

4.2   LLVM coding style

LLVM adheres to specific code-style rules [11]. The primary objective of these rules is to promote proficient C++ practices with a special focus on performance. As previously mentioned, LLVM employs C++17 and prefers using data structures and algorithms from the **STL** (short for, **Standard Template Library**). On the other hand, LLVM offers many optimized versions of data structures that mirror those in the STL. For example, `llvm::SmallVector<>` can be regarded as an optimized version of `std::vector<>`, especially for small sizes of the vector, a common trait for data structures used in compilers.

Given a choice between an STL object/algorithm and its corresponding LLVM version, the LLVM coding standard advises favoring the LLVM version.

Additional rules pertain to concerns regarding performance limitations. For instance, both **run-time type information (RTTI)** and C++ exceptions are disallowed. However, there are situations where RTTI could prove beneficial; thus, LLVM offers alternatives such as `llvm::isa<>` and other similar template helper functions. More information on this can be found in *Section 4.3.1, RTTI replacement and cast operators*. Instead of C++ exceptions,

LLVM frequently employs C-style `asserts`.

Sometimes, asserts are not sufficiently informative. LLVM recommends adding textual messages to them to simplify debugging. Here's a typical example from Clang's code:

```
static bool unionHasUniqueObjectRepresentations(const ASTContext &Context,
                                                const RecordDecl *RD,
                                                bool
                                          ↪ CheckIfTriviallyCopyable)
                                          ↪ {
  assert(RD->isUnion() && "Must be union type");
  CharUnits UnionSize = Context.getTypeSizeInChars(RD->getTypeForDecl());
```

Figure 4.1: Usage of assert() in clang/lib/AST/ASTContext.cpp

In the code, we check if the second parameter (RD) is a union and raise an assert with a corresponding message if it's not.

Besides performance considerations, LLVM also introduces some additional requirements. One of these requirements concerns comments. Code comments are very important. Furthermore, both LLVM and Clang have comprehensive documentation generated from the code. They use Doxygen (`https://www.doxygen.nl/`) for this purpose. This tool is the de facto standard for commenting in C/C++ programs, and you have most likely encountered it before.

Clang and LLVM are not monolithic pieces of code; instead, they are implemented as a set of libraries. This design provides advantages in terms of code and functionality reuse, as we will explore in *Chapter 8, IDE Support and Clangd*. These libraries also serve as excellent examples of LLVM code style enforcement. Let's examine these libraries in detail.

4.3   LLVM basic libraries

We are going to start with RTTI replacement in the LLVM code and discuss how it's implemented. We will then continue with basic containers and smart pointers. We will conclude with some important classes used to represent token locations and how diagnostics are realized in Clang. Later, in *Section 4.6, Clang plugin project*, we will use some of these classes in our test project.

4.3.1   RTTI replacement and cast operators

As mentioned earlier, LLVM avoids using RTTI due to performance concerns. LLVM has introduced several helper functions that replace RTTI counterparts, allowing for the casting of an object from one type to another. The fundamental ones are as follows:

- `llvm::isa<>` is akin to Java's `instanceof` operator. It returns `true` or `false` depending on whether the reference to the tested object belongs to the tested class or not.

- `llvm::cast<>`: Use this cast operator when you're certain that the object is of the specified derived type. If the cast fails (i.e., the object isn't of the expected type), `llvm::cast` will abort the program. Use it only when you're confident the cast won't fail.

- `llvm::dyn_cast<>`: This is perhaps the most frequently used casting operator in LLVM. `llvm::dyn_cast` is employed for safe downcasting when you anticipate the cast will usually succeed, but there's some uncertainty. If the object isn't of the specified derived type, `llvm::dyn_cast<>` returns `nullptr`.

The cast operators do not accept `nullptr` as input. However, there are two special cast operators that can handle null pointers:

- `llvm::cast_if_present<>`: A variant of `llvm::cast<>` that accepts `nullptr` values

- `llvm::dyn_cast_if_present<>`: A variant of `llvm::dyn_cast<>` that accepts `nullptr` values

Both operators can handle `nullptr` values. If the input is `nullptr` or if the cast fails, they simply return `nullptr`.

Important note

It's worth noting that the casting operators `llvm::cast_if_present<>` and `llvm::dyn_cast_if_present<>` were introduced recently, specifically in 2022. They serve as replacements for popular ones `llvm::cast_or_null<>` and `llvm::dyn_cast_or_null<>`, which had been in recent use. The older versions are still supported and now redirect calls to the newer cast operators. For more information, see the discussion about this change: `https://discourse.llvm.org/t/psa-swapping-out-or-null-with-if-present/65018`

The following question might arise: how can the dynamic cast operation be performed without RTTI? This can be achieved with certain specific decorations, as illustrated in a simple example inspired by *How to set up LLVM-style RTTI for your class hierarchy* [14]. We'll begin with a base class, `clangbook::Animal`, that has two descendants: `clangbook::Horse` and `clangbook::Sheep`. Each horse can be categorized by its speed (in mph), and each sheep by its wool mass. Here's how it can be used:

```
46 void testAnimal() {
47   auto AnimalPtr = std::make_unique<clangbook::Horse>(10);
48   if (llvm::isa<clangbook::Horse>(AnimalPtr)) {
49     llvm::outs()
50         << "Animal is a Horse and the horse speed is: "
51         << llvm::dyn_cast<clangbook::Horse>(AnimalPtr.get())->getSpeed()
52         << "mph \n";
53   } else {
54     llvm::outs() << "Animal is not a Horse\n";
55   }
56 }
```

Figure 4.2: LLVM `isa<>` *and* `dyn_cast<>` *usage example*

The code should produce the following output:

```
Animal is a Horse and the horse speed is: 10mph
```

Line 48 in Figure 4.2 demonstrates the use of llvm::isa<>, while *Line 51* showcases llvm::dyn_cast<>. In the latter, we cast the base class to clangbook::Horse and call a method specific to that class.

Let's look into the class implementations, which will provide insights into how the RTTI replacement works. We will start with the base class clangbook::Animal:

```
 9 class Animal {
10 public:
11   enum AnimalKind { AK_Horse, AK_Sheep };
12
13 public:
14   Animal(AnimalKind K) : Kind(K){};
15   AnimalKind getKind() const { return Kind; }
16
17 private:
18   const AnimalKind Kind;
19 };
```

Figure 4.3: clangbook::Animal *class*

The most crucial aspect is *Line 11* in the preceding code. It specifies different "kinds" of animals. One enum value is used for the horse (AK_Horse) and another for the sheep (AK_Sheep). Hence, the base class has some knowledge about its descendants. The implementations for the clangbook::Horse and clangbook::Sheep classes can be found in the following code:

```
21 class Horse : public Animal {
22 public:
23   Horse(int S) : Animal(AK_Horse), Speed(S){};
24
25   static bool classof(const Animal *A) { return A->getKind() == AK_Horse; }
26
27   int getSpeed() { return Speed; }
28
29 private:
30   int Speed;
31 };
32
33 class Sheep : public Animal {
34 public:
35   Sheep(int WM) : Animal(AK_Sheep), WoolMass(WM){};
36
37   static bool classof(const Animal *A) { return A->getKind() == AK_Sheep; }
38
39   int getWoolMass() { return WoolMass; }
40
41 private:
42   int WoolMass;
43 };
```

Figure 4.4: clangbook::Horse *and* clangbook::Sheep *classes*

Lines 25 and 37 are particularly important as they contain the classof static method implementation. This method is crucial for the cast operators in LLVM. A typical implementation might look like the following (simplified version):

```
1 template <typename To, typename From>
2 bool isa(const From *Val) {
3   return To::classof(Val);
4 }
```

Figure 4.5: Simplified implementation for llvm::isa<>

The same mechanism can be applied to other cast operators.

Our next topic will discuss various types of containers that serve as more powerful alternatives to their corresponding STL counterparts.

4.3.2   Containers

The LLVM ADT (which stands for Abstract Data Type) library offers a set of containers. While some of them are unique to LLVM, others can be considered as replacements for containers from the STL. We will explore some of the most popular classes provided by the ADT.

String operations

The primary class for working with strings in the standard C++ library is std::string. Although this class was designed to be universal, it has some performance related issues. A significant issue concerns the copy operation. Since copying strings is a common operation in compilers, LLVM introduced a specialized class, llvm::StringRef, that handles this operation efficiently without using extra memory. This class is comparable to std::string_view from C++17 [20] and std::span from C++20 [21].

The llvm::StringRef class maintains a reference to data, which doesn't need to be null-terminated like traditional C/C++ strings. It essentially holds a pointer to a data block and the block's size, making the object's effective size 16 bytes. Because llvm::StringRef retains a reference rather than the actual data, it must be constructed from an existing data source. This class can be instantiated from basic string objects such as **const char***, std::string, and std::string_view. The default constructor creates an empty object.

Typical usage example for llvm::StringRef is shown in Figure 4.6:

```
1   #include "llvm/ADT/StringRef.h"
2   ...
3   llvm::StringRef StrRef("Hello, LLVM!");
4   // Efficient substring, no allocations
5   llvm::StringRef SubStr = StrRef.substr(0, 5);
6
7   llvm::outs() << "Original StringRef: " << StrRef.str() << "\n";
8   llvm::outs() << "Substring: " << SubStr.str() << "\n";
```

Figure 4.6: llvm::StringRef *usage example*

The output for the code is shown here:

```
Original StringRef: Hello, LLVM!
Substring: Hello
```

Another class used for string manipulation in LLVM is llvm::Twine, which is particularly useful when concatenating several objects into one. A typical usage example for the class is shown in Figure 4.7:

```
1   #include "llvm/ADT/Twine.h"
2   ...
3   llvm::StringRef Part1("Hello, ");
4   llvm::StringRef Part2("Twine!");
5   llvm::Twine Twine = Part1 + Part2;   // Efficient concatenation
6
7   // Convert twine to a string (actual allocation happens here)
8   std::string TwineStr = Twine.str();
9   llvm::outs() << "Twine result: " << TwineStr << "\n";
```

Figure 4.7: llvm::Twine *usage example*

The output for the code is shown here:

```
Twine result: Hello, Twine!
```

Another class that is widely used for string manipulations is llvm::SmallString<>. It represents a string that is stack-allocated up to a fixed size, but can also grow beyond this size, at which point it heap-allocates memory. This is a blend between the space efficiency of stack allocation and the flexibility of heap allocation.

The advantage of llvm::SmallString<> is that for many scenarios, especially in compiler tasks, strings tend to be small and fit within the stack-allocated space. This avoids the overhead of dynamic memory allocation. But in situations where a larger string is required, llvm::SmallString can still accommodate by transitioning to heap memory. A typical usage example is show in Figure 4.8:

```
1    #include "llvm/ADT/SmallString.h"
2    ...
3    // Stack allocate space for up to 20 characters.
4    llvm::SmallString<20> SmallStr;
5
6    // No heap allocation happens here.
7    SmallStr = "Hello, ";
8    SmallStr += "LLVM!";
9
10   llvm::outs() << "SmallString result: " << SmallStr << "\n";
```

Figure 4.8: llvm::SmallString<> *usage example*

Despite the fact that string manipulation is key in compiler tasks such as text parsing, LLVM has many other helper classes. We'll explore its sequential containers next.

Sequential containers

LLVM recommends some optimized replacements for arrays and vectors from the standard library. The most notable are:

- llvm::ArrayRef<>: A helper class designed for interfaces that accept a sequential list of elements for read-only access. The class is akin to llvm::StringRef<> in that it does not own the underlying data but merely references it.

- llvm::SmallVector<>: An optimized vector for cases with a small size. It resembles llvm::SmallString, as discussed in *Section 4.3.2, String operations*. Notably, the size for the array isn't fixed, allowing the number of stored elements to grow. If the number of elements stays below N (the template argument), then there is no need for additional memory allocation.

Let's examine the llvm::SmallVector<> to better understand these containers, as shown in Figure 4.9:

```
1   llvm::SmallVector<int, 10> SmallVector;
2   for (int i = 0; i < 10; i++) {
3     SmallVector.push_back(i);
4   }
5   SmallVector.push_back(10);
```

Figure 4.9: llvm::SmallVector<> *usage*

The vector is initialized at *Line 1* with a chosen size of 10 (indicated by the second template argument). The container offers an API similar to std::vector<>, using the familiar push_back method to add new elements, as seen in Figure 4.9, *Lines 3 and 5*.

The first 10 elements are added to the vector without any additional memory allocation (see Figure 4.9, *Lines 2-4*). However, when the eleventh element is added at *Line 5*, the array's size surpasses the pre-allocated space for 10 elements, triggering additional memory allocation. This container design efficiently minimizes memory allocation for small objects while maintaining the flexibility to accommodate larger sizes when necessary.

Map-like containers

The standard library provides several containers for storing key-value data. The most common ones are `std::map<>` for general-purpose maps and `std::unordered_map<>` for hash maps. LLVM offers additional alternatives to these standard containers:

- `llvm::StringMap<>`: A map that uses strings as keys. Typically, this is more performance optimized than the standard associative container, `std::unordered_map<std::string, T>`. It is frequently used in situations where string keys are dominant and performance is critical, as one might expect in a compiler infrastructure like LLVM. Unlike many other data structures in LLVM, `llvm::StringMap<>` does not store a copy of the string key. Instead, it keeps a reference to the string data, so it's crucial to ensure the string data outlives the map to prevent undefined behavior.

- `llvm::DenseMap<>`: This map is designed to be more memory- and time-efficient than `std::unordered_map<>` in most situations, though it comes with some additional constraints (e.g., keys and values having trivial destructors). It's especially beneficial when you have simple key-value types and require high-performance lookups.

- `llvm::SmallDenseMap<>`: This map is akin to `llvm::DenseMap<>` but is optimized for instances where the map size is typically small. It allocates from the stack for small maps and only resorts to heap allocation when the map exceeds a predefined size.

- `llvm::MapVector<>`: This container retains the insertion order, akin to Python's `OrderedDict`. It is implemented as a blend of `std::vector` and either `llvm::DenseMap` or `llvm::SmallDenseMap`.

It's noteworthy that these containers utilize a quadratically probed hash table mechanism. This method is effective for hash collision resolution because the cache isn't recomputed during element lookups. This is crucial for performance-critical applications, such as compilers.

4.3.3 Smart pointers

Different smart pointers can be found in LLVM code. The most popular ones come from the standard template library: `std::unique_ptr<>` and `std::shared_ptr<>`. In addition, LLVM provides some supplementary classes to work with smart pointers. One of the most prominent among them is `llvm::IntrusiveRefCntPtr<>`. This smart pointer is designed to work with objects that support intrusive reference counting. Unlike `std::shared_ptr`, which maintains its own control block to manage the reference count, `IntrusiveRefCntPtr` expects the object to maintain its own reference count. This design can be more memory efficient. A typical usage example is shown here:

```
1   class MyClass : public llvm::RefCountedBase<MyClass> {
2   // ...
3   };
4
5   llvm::IntrusiveRefCntPtr<MyClass> Ptr = new MyClass();
```

Figure 4.10: `llvm::IntrusiveRefCntPtr<>` *usage example*

As we can see, the smart pointer prominently employs the CRTP (which stands for Curiously Recurring Template Pattern) that was mentioned earlier in *Section 3.3, AST traversal*. The CRTP is essential for the `Release` operation when the reference count drops to 0 and the object must be deleted. The implementation is as follows:

```
1 template <class Derived> class RefCountedBase {
2   // ...
3   void Release() const {
4     assert(RefCount > 0 && "Reference count is already zero.");
5     if (--RefCount == 0)
6       delete static_cast<const Derived *>(this);
7   }
8 }
```

Figure 4.11: CRTP usage in llvm::RefCountedBase<>. *The code was sourced from the* llvm/ADT/IntrusiveRefCntPtr.h *header*

Since MyClass in Figure 4.10 is derived from RefCountedBase, we can perform a cast on it in *Line 6* of Figure 4.11. This cast is feasible since the type to cast is known, given that it is provided as a template parameter.

We've just finished with LLVM basic libraries. Now it is time to move on to Clang basic libraries. Clang is a compiler frontend, and its most important operations are related to diagnostics. Diagnostics require precise information about position location in the source code. Let's explore the basic classes that Clang provides for these operations.

4.4   Clang basic libraries

Clang is a compiler frontend, and its most important operations are related to diagnostics. Diagnostics require precise information about position location in the source code. Let's explore the basic classes that Clang provides for these operations.

4.4.1   SourceManager and SourceLocation

Clang, as a compiler, operates with text files (programs), and locating a specific place in the program is one of the most frequently requested operations. Let's look at a typical Clang error report. Consider a program from *Chapter 3, Clang AST*, as seen in Figure 3.33. Clang produces the following error message for the program:

```
$ <...>/llvm-project/install/bin/clang -fsyntax-only maxerr.cpp
maxerr.cpp:3:12: error: use of undeclared identifier 'ab'
    return ab;
              ^
1 error generated.
```

Figure 4.12: Error reported in maxerr.cpp

As we can see in Figure 4.12, the following information is required to display the message:

- Filename: In our case, it's `maxerr.cpp`

- Line in the file: In our case, it's 3

- Column in the file: In our case, it's 12

The data structure that stores this information should be as compact as possible because the compiler uses it frequently. Clang stores the required information in the `clang::SourceLocation` object.

This object is used often, so it should be small in size and quick to copy. We can check the size of the object using lldb. For instance, if we run Clang under the debugger, we can determine the size as follows:

```
$ lldb <...>/llvm-project/install/clang
...
(lldb) p sizeof(clang::SourceLocation)
(unsigned long) 4
(lldb)
```

Figure 4.13: clang::SourceLocation size determination under debugger

That is, the information is encoded using a single `unsigned long` number. How is this possible? The number merely serves as an identifier for a position in the text file. An additional class is required to correctly extract and represent this information, which is clang::SourceManager. The SourceManager object contains all the details about a specific location. In Clang, managing source locations can be challenging due to the presence of macros, includes, and other preprocessing directives. Consequently, there are several ways to interpret a given source location. The primary ones are as follows:

- **Spelling location**: Refers to the location where something was actually spelled out in the source. If you have a source location pointing inside a macro body, the spelling location will give you the location in the source code where the contents of the macro are defined.

- **Expansion location**: Refers to where a macro gets expanded. If you have a source location pointing inside a macro body, the expansion location will give you the location in the source code where the macro was used (expanded).

Let's look at a specific example:

```
1 #define BAR void bar()
2 int foo(int x);
3 BAR;
```

Figure 4.14: Example program to test different types of source locations: functions.hpp

In Figure 4.14, we define two functions: int foo() at *Line 2* and void bar() at *Line 3*. For the first function, both the spelling and expansion locations point to *Line 2*. However, for the second function, the spelling location is at *Line 1*, while the expansion location is at *Line 3*.

Let's examine this with a test Clang tool. We will use the test project from *Section 3.4, Recursive AST visitor* and replace some parts of the code here. First of all, we have to pass clang::ASTContext to our Visitor implementation. This is required because clang::ASTContext provides access to clang::SourceManager. We will replace *Line 11* in Figure 3.8 and pass ASTContext as follows:

```
10   CreateASTConsumer(clang::CompilerInstance &CI, llvm::StringRef File) {
11     return std::make_unique<Consumer>(&CI.getASTContext());
```

The Consumer class (see Figure 3.9) will accept the argument and use it as a parameter for Visitor:

```
8   Consumer(clang::ASTContext *Context)
9       : V(std::make_unique<Visitor>(Context)) {}
```

The main changes are for the Visitor class, which is mostly rewritten. First of all, we pass clang::ASTContext to the class constructor as follows:

```
5 class Visitor : public clang::RecursiveASTVisitor<Visitor> {
6 public:
7   explicit Visitor(clang::ASTContext *C) : Context(C) {}
8
```

Figure 4.15: Visitor class implementation: constructor

The AST Context class is stored as a private member of our class, as shown below:

```
25 private:
26   clang::ASTContext *Context;
```

Figure 4.16: Visitor class implementation: private section

The main processing logic is in Visitor::VisitFunctionDecl method, which you can see next

```
9    bool VisitFunctionDecl(const clang::FunctionDecl *FD) {
10     clang::SourceManager &SM = Context->getSourceManager();
11     clang::SourceLocation Loc = FD->getLocation();
12     clang::SourceLocation ExpLoc = SM.getExpansionLoc(Loc);
13     clang::SourceLocation SpellLoc = SM.getSpellingLoc(Loc);
14     llvm::StringRef ExpFileName = SM.getFilename(ExpLoc);
15     llvm::StringRef SpellFileName = SM.getFilename(SpellLoc);
16     unsigned SpellLine = SM.getSpellingLineNumber(SpellLoc);
17     unsigned ExpLine = SM.getExpansionLineNumber(ExpLoc);
18     llvm::outs() << "Spelling : " << FD->getName() << " at " <<
        ↪ SpellFileName
19                 << ":" << SpellLine << "\n";
20     llvm::outs() << "Expansion : " << FD->getName() << " at " <<
        ↪ ExpFileName
21                 << ":" << ExpLine << "\n";
22     return true;
23   }
```

Figure 4.17: Visitor class implementation: VisitFunctionDecl method

If we compile and run the code on the test file from Figure 4.14, the following output will
be generated::

```
Spelling : foo at functions.hpp:2
Expansion : foo at functions.hpp:2
Spelling : bar at functions.hpp:1
Expansion : bar at functions.hpp:3
```

Figure 4.18: Output from the recursivevisitor executable on the functions.hpp test file

clang::SourceLocation and clang::SourceManager are very powerful classes. In
combination with other classes such as clang::SourceRange (a pair of two source locations
that specify the beginning and end of a source range), they provide a great foundation for

diagnostics used in Clang.

4.4.2 Diagnostics support

Clang's diagnostics subsystem is responsible for generating and reporting warnings, errors, and other messages [8]. The main classes involved are:

- DiagnosticsEngine: Manages diagnostic IDs and options

- DiagnosticConsumer: Abstract base class for diagnostic consumers

- DiagnosticIDs: Handles the mapping between diagnostic flags and internal IDs

- DiagnosticInfo: Represents a single diagnostic

Here is a simple example illustrating how you might emit a warning in Clang:

```
18  // Emit a warning
19  DiagnosticsEngine.Report(DiagnosticsEngine.getCustomDiagID(
20      clang::DiagnosticsEngine::Warning, "This is a custom warning."));
```

Figure 4.19: Emit warning with clang::DiagnosticsEngine

In our example, we will use a simple DiagnosticConsumer, clang::TextDiagnosticPrinter, which formats and prints the processed diagnostic messages.

The full code for the main function of our example is shown in Figure 4.20:

```
 7 int main() {
 8    llvm::IntrusiveRefCntPtr<clang::DiagnosticOptions> DiagnosticOptions =
 9        new clang::DiagnosticOptions();
10    clang::TextDiagnosticPrinter TextDiagnosticPrinter(
11        llvm::errs(), DiagnosticOptions.get(), false);
12
13    llvm::IntrusiveRefCntPtr<clang::DiagnosticIDs> DiagIDs =
14        new clang::DiagnosticIDs();
15    clang::DiagnosticsEngine DiagnosticsEngine(DiagIDs, DiagnosticOptions,
16                                               &TextDiagnosticPrinter,
                                             ↪  false);
17
18    // Emit a warning
19    DiagnosticsEngine.Report(DiagnosticsEngine.getCustomDiagID(
20        clang::DiagnosticsEngine::Warning, "This is a custom warning."));
21
22    return 0;
23 }
```

Figure 4.20: Clang diagnostics example

The code will produce the following output

```
warning: This is a custom warning.
```

Figure 4.21: Printed diagnostics

In this example, we first set up DiagnosticsEngine with TextDiagnosticPrinter as its DiagnosticConsumer. We then use the Report method of DiagnosticsEngine to emit a custom warning. We will add a more realistic example later when we create our test project for the Clang plugin in *Section 4.6, Clang plugin project*.

4.5   LLVM supporting tools

The LLVM project has its own tooling support. The most important LLVM tools are TableGen and LIT (which stands for LLVM Integrated Tester). We will look into them with examples from the Clang code. These examples should help us understand the purpose of the tooling and how they can be used.

4.5.1   TableGen

TableGen is a **domain-specific language (DSL)** and associated tool used in the LLVM project for the purpose of describing and generating tables, particularly those that describe a target architecture. This is highly useful for compiler infrastructure, where one frequently needs to describe things such as instruction sets, registers, and various other target-specific attributes in a structured manner.

TableGen is employed in various parts of the Clang compiler. It's primarily used where there's a need to generate large amounts of similar code. For instance, it can be used for supporting cast operations that necessitate extensive enum declarations in basic classes, or in the diagnostic subsystem where code generation is required to handle numerous similar diagnostic messages. We will examine how TableGen functions within the diagnostics system as an example.

We will begin with the `Diagnostic.td` file, which describes Clang's diagnostics. This file can be found at `clang/include/clang/Basic/Diagnostic.td`. Let's examine how diagnostic severity is defined:

```
16  // Define the diagnostic severities.
17  class Severity<string N> {
18    string Name = N;
19  }
```

Figure 4.22: Severity definition in clang/include/clang/Basic/Diagnostic.td

In Figure 4.22, we define a class for severities (*Lines 17-19*). Each severity is associated with a string, as shown below:

```
20 def SEV_Ignored : Severity<"Ignored">;
21 def SEV_Remark  : Severity<"Remark">;
22 def SEV_Warning : Severity<"Warning">;
23 def SEV_Error   : Severity<"Error">;
24 def SEV_Fatal   : Severity<"Fatal">;
```

Figure 4.23: Definitions for different types of severity in clang/include/clang/Basic/Diagnostic.td

Figure 4.23 contains definitions for the different severities; for instance, the Warning severity is defined on *Line 22*.

The severity is later used to define the Diagnostic class, with the Warning diagnostic being defined as a descendant of this class:

```
// All diagnostics emitted by the compiler are an indirect subclass of
↪   this.
class Diagnostic<string summary, DiagClass DC, Severity defaultmapping> {
  ...
}
...
class Warning<string str>   : Diagnostic<str, CLASS_WARNING, SEV_Warning>;
```

Figure 4.24: Diagnostics definition in clang/include/clang/Basic/Diagnostic.td

Using the Warning class definition, different instances of the class can be defined. For example, the following is an instance that defines an unused parameter warning located in DiagnosticSemaKinds.td:

```
def warn_unused_parameter : Warning<"unused parameter %0">,
  InGroup<UnusedParameter>, DefaultIgnore;
```

Figure 4.25: The definition of the unused parameter warning in
clang/include/clang/Basic/DiagnosticSemaKinds.td

The `clang-tblgen` tool will generate the corresponding `DiagnosticSemaKinds.inc` file:

```
DIAG(warn_unused_parameter, CLASS_WARNING,
↪  (unsigned)diag::Severity::Ignored, "unused parameter %0", 985,
↪  SFINAE_Suppress, false, false, true, false, 2)
```

Figure 4.26: The definition of the unused parameter warning in
clang/include/clang/Basic/DiagnosticSemaKinds.inc

This file retains all the necessary information about the diagnostic. This information can be retrieved from the Clang source code using different definitions of the `DIAG` macro.

For instance, the following code leverages the TableGen-generated code to extract diagnostic descriptions, as found in `clang/lib/Basic/DiagnosticIDs.cpp`:

```
const StaticDiagInfoDescriptionStringTable StaticDiagInfoDescriptions = {
#define DIAG(ENUM, CLASS, DEFAULT_SEVERITY, DESC, GROUP, SFINAE, NOWERROR,\
             SHOWINSYSHEADER, SHOWINSYSMACRO, DEFERRABLE, CATEGORY)        \
  DESC,
...
#include "clang/Basic/DiagnosticSemaKinds.inc"
...
#undef DIAG
};
```

Figure 4.27: DIAG macro definition

The C++ preprocessor will expand to the following:

```
const StaticDiagInfoDescriptionStringTable StaticDiagInfoDescriptions = {

  ...

  "unused parameter %0",

  ...

};
```

Figure 4.28: DIAG macro expansion

The provided example demonstrates how TableGen can be used to generate code in Clang and how it can simplify Clang development. The diagnostic subsystem is not the only area where TableGen is utilized; it is also widely used in other parts of Clang. For instance, the macros used in various types of AST visitors also rely on the code generated by TableGen; see *Section 3.3.2, Visitor implementation*.

4.5.2   LLVM test framework

LLVM uses several testing frameworks for different types of testing. The primary ones are **LLVM Integrated Tester (LIT)** and **Google Test (GTest)** [24]. Both LIT and GTest play significant roles in Clang's testing infrastructure:

- LIT is primarily used for testing the behavior of the Clang toolchain as a whole, with a focus on its code compilation capabilities and the diagnostics it produces.

- GTest is utilized for unit tests, targeting specific components of the code base, primarily utility libraries and internal data structures.

These tests are crucial for maintaining the quality and stability of the Clang project.

> **Important note**
>
> We will not delve into GTest, as this testing framework is commonly used outside LLVM and isn't part of LLVM itself. For more information about GTest, please visit its official page: `https://github.com/google/googletest`

Our focus will be on LIT. LIT is LLVM's own test framework and is heavily used for testing the various tools and libraries in LLVM, including the Clang compiler. LIT is designed to be lightweight and is tailored for the needs of compiler testing. It's commonly used for running tests that are essentially shell scripts, often with checks for specific patterns in the output. A typical LIT test may consist of a source code file along with a set of "RUN" commands that specify how to compile, link, or otherwise process the file, and what output to expect.

The RUN commands often use FileCheck, another utility in the LLVM project, to check the output against expected patterns. In Clang, LIT tests are often used to test frontend features such as parsing, semantic analysis, code generation, and diagnostics. These tests typically look like source code files with embedded comments to indicate how to run the test and what to expect.

Consider the following example from `clang/test/Sema/attr-unknown.c`:

```
1 // RUN: %clang_cc1 -fsyntax-only -verify -Wattributes %s
2
3 int x __attribute__((foobar)); // expected-warning {{unknown attribute
  ↳  'foobar' ignored}}
4 void z(void) __attribute__((bogusattr)); // expected-warning {{unknown
  ↳  attribute 'bogusattr' ignored}}
```

Figure 4.29: LIT test for Clang warnings about unknown attributes

The example is a typical C source file that can be processed by Clang. LIT's behavior is controlled by comments within the source text. The first comment (on *Line 1*) specifies how the test should be executed. As indicated, `clang` should be started with some additional arguments: `-fsyntax-only` and `-verify`. There are also substitutions that begin with the '%' symbol. The most important of these is '%s', which is replaced by the source file's name. LIT will also examine comments beginning with `expected-warning` and ensure that the warnings produced by Clang's output match the expected values.

The test can be run as follows:

```
$ ./build/bin/llvm-lit ./clang/test/Sema/attr-unknown.c

...

-- Testing: 1 tests, 1 workers --
PASS: Clang :: Sema/attr-unknown.c (1 of 1)


Testing Time: 0.06s
  Passed: 1
```

Figure 4.30: LIT test run

We run `llvm-lit` from the `build` folder because the tool is not included in the installation procedure. We can obtain more details about LIT setup and its invocation once we create our test clang plugin project and configure LIT tests for it.

4.6   Clang plugin project

The goal of the test project is to create a clang plugin that will estimate class complexity. Specifically, a class is deemed complex if the number of its methods exceeds a certain threshold. We will leverage all the knowledge we have acquired thus far for this project. This will include the use of a recursive visitor and Clang diagnostics. Additionally, we will create a LIT test for our project. Developing the plugin will necessitate a unique build configuration for LLVM, which will be our initial step.

4.6.1   Environment setup

The plugin will be created as a shared object, and our LLVM installation should be built with support for shared libraries (see *Section 1.3.1, Configuration with CMake*):

```
cmake -G Ninja -DCMAKE_BUILD_TYPE=Debug -DCMAKE_INSTALL_PREFIX=../install
  ↳  -DLLVM_TARGETS_TO_BUILD="X86" -DLLVM_ENABLE_PROJECTS="clang"
  ↳  -DLLVM_USE_SPLIT_DWARF=ON -DBUILD_SHARED_LIBS=ON ../llvm
```

Figure 4.31: CMake configuration used for the Clang plugin project

As can be seen, we use the build configuration from *Section 1.4, Test project – syntax check with a Clang tool*, as shown in Figure 1.12. In the configuration, we set up a folder for installing artifacts into ../install, limit our build targets to the X86 platform, and enable only the clang project. Additionally, we enable size optimization for debug symbols and use shared libraries instead of static linkage.

The next step involves building and installing clang. This can be achieved with the following command:

```
$ ninja install
```

As soon as we are done with the clang build and installation, we can proceed with the CMakeLists.txt file for our project.

4.6.2 CMake build configuration for plugin

We will use Figure 3.20 as the foundation for our plugin build configuration. We will change the project name to classchecker , and ClassComplexityChecker.cpp will serve as our primary source file. The main portion of the file is displayed in Figure 4.32. As can be observed, we will construct a shared library (*Lines 18-20*) rather than an executable, as in our previous test projects. Another modification is in *Line 12*, where we set up a config parameter for the LLVM build folder. This parameter is necessary to locate the LIT executable, which is not included in the standard installation process, as mentioned earlier in *Section 4.5.2, LLVM test framework*. Some additional modifications need to be made to support LIT test invocations, but we will discuss the details later in *Section 4.6.8, LIT tests for clang plugin* (see Figure 4.44).

```
 8    message(STATUS "$LLVM_HOME found: $ENV{LLVM_HOME}")
 9    set(LLVM_HOME $ENV{LLVM_HOME} CACHE PATH "Root of LLVM installation")
10    set(LLVM_LIB ${LLVM_HOME}/lib)
11    set(LLVM_DIR ${LLVM_LIB}/cmake/llvm)
12    set(LLVM_BUILD $ENV{LLVM_BUILD} CACHE PATH "Root of LLVM build")
13    find_package(LLVM REQUIRED CONFIG)
14    include_directories(${LLVM_INCLUDE_DIRS})
15    link_directories(${LLVM_LIBRARY_DIRS})
16
17    # Add the plugin's shared library target
18    add_library(classchecker MODULE
19      ClassChecker.cpp
20    )
21    set_target_properties(classchecker PROPERTIES COMPILE_FLAGS "-fno-rtti")
22    target_link_libraries(classchecker
23      LLVMSupport
24      clangAST
25      clangBasic
26      clangFrontend
27      clangTooling
28    )
```

Figure 4.32: CMakeLists.txt file for class complexity plugin

After completing the build configuration, we can start writing the primary code for the plugin. The first component we'll create is a recursive visitor class named ClassVisitor.

4.6.3 Recursive visitor class

Our visitor class is located in the ClassVisitor.hpp file (see Figure 4.33). This is a recursive visitor that handles clang::CXXRecordDecl, which are the AST nodes for C++ class declarations. We calculate the number of methods in *Lines 13-16* and emit diagnostics in *Lines 19-25* if the threshold is exceeded.

```
1 #include "clang/AST/ASTContext.h"
2 #include "clang/AST/RecursiveASTVisitor.h"
3
4 namespace clangbook {
5 namespace classchecker {
6 class ClassVisitor : public clang::RecursiveASTVisitor<ClassVisitor> {
7 public:
8   explicit ClassVisitor(clang::ASTContext *C, int T)
9       : Context(C), Threshold(T) {}
10
11   bool VisitCXXRecordDecl(clang::CXXRecordDecl *Declaration) {
12     if (Declaration->isThisDeclarationADefinition()) {
13       int MethodCount = 0;
14       for (const auto *M : Declaration->methods()) {
15         MethodCount++;
16       }
17
18       if (MethodCount > Threshold) {
19         clang::DiagnosticsEngine &D = Context->getDiagnostics();
20         unsigned DiagID =
21             D.getCustomDiagID(clang::DiagnosticsEngine::Warning,
22                               "class %0 is too complex: method count =
                                 ↪ %1");
23         clang::DiagnosticBuilder DiagBuilder =
24             D.Report(Declaration->getLocation(), DiagID);
25         DiagBuilder << Declaration->getName() << MethodCount;
26       }
27     }
28     return true;
29   }
```

```
30
31 private:
32   clang::ASTContext *Context;
33   int Threshold;
34 };
35 } // namespace classchecker
36 } // namespace clangbook
```

Figure 4.33: Source code for ClassVisitor.hpp

It's worth noting the diagnostic calls. The diagnostic message is constructed in *Lines 20-22*. Our diagnostic message accepts two parameters: the class name and the number of methods for the class. These parameters are encoded with the '%1' and '%2' placeholders in *Line 22*. The actual values for these parameters are passed in *Line 25*, where the diagnostic message is constructed using the `DiagBuild` object. This object is an instance of the `clang::DiagnosticBuilder` class, which implements the **Resource Acquisition Is Initialization (RAII)** pattern. It emits the actual diagnostics upon its destruction.

> **Important note**
>
> In C++, the RAII principle is a common idiom used to manage resource lifetimes by tying them to the lifetime of an object. When an object goes out of scope, its destructor is automatically called, and this provides an opportunity to release the resource that the object holds.

`ClassVisitor` is created within an AST consumer class, which will be our next topic.

4.6.4   Plugin AST consumer class

The AST consumer class is implemented in `ClassConsumer.hpp` and represents the standard AST consumer, as seen in our AST visitor test projects (refer to Figure 3.9). The code is presented in Figure 4.35.

```
 5 namespace clangbook {
 6 namespace classchecker {
 7 class ClassConsumer : public clang::ASTConsumer {
 8 public:
 9   explicit ClassConsumer(clang::ASTContext *Context, int Threshold)
10       : Visitor(Context, Threshold) {}
11
12   virtual void HandleTranslationUnit(clang::ASTContext &Context) {
13     Visitor.TraverseDecl(Context.getTranslationUnitDecl());
14   }
15
16 private:
17   ClassVisitor Visitor;
18 };
19 } // namespace classchecker
20 } // namespace clangbook
```

Figure 4.34: Source code for ClassConsumer.hpp

The code initializes Visitor at *Line 10* and utilizes the Visitor class at *Line 13* to traverse the declarations, starting with the top one (translation unit declaration). The consumer must be created from a special AST action class, which we will discuss next.

4.6.5 Plugin AST action class

The code for the AST action is shown in Figure 4.35. Several important parts can be observed:

- *Line 7*: We inherit our ClassAction from clang::PluginASTAction

- *Lines 10-13*: We instantiate ClassConsumer and utilize MethodCountThreshold, which is derived from an optional plugin argument

- *Lines 15-25*: We process the optional threshold argument for our plugin

```
5 namespace clangbook {
6 namespace classchecker {
7 class ClassAction : public clang::PluginASTAction {
8 protected:
9   std::unique_ptr<clang::ASTConsumer>
10  CreateASTConsumer(clang::CompilerInstance &CI, llvm::StringRef) {
11    return std::make_unique<ClassConsumer>(&CI.getASTContext(),
12                                           MethodCountThreshold);
13  }
14
15  bool ParseArgs(const clang::CompilerInstance &CI,
16                 const std::vector<std::string> &args) {
17    for (const auto &arg : args) {
18      if (arg.substr(0, 9) == "threshold") {
19        auto valueStr = arg.substr(10); // Get the substring after
            ↪ "threshold="
20        MethodCountThreshold = std::stoi(valueStr);
21        return true;
22      }
23    }
24    return true;
25  }
26  ActionType getActionType() { return AddAfterMainAction; }
27
28 private:
29  int MethodCountThreshold = 5; // default value
30 };
31 } // namespace classchecker
32 } // namespace clangbook
```

Figure 4.35: Source code for ClassAction.hpp

We are almost done and ready to initialize our plugin.

4.6.6 Plugin code

Our plugin registration is carried out in the `ClassChecker.cpp` file, shown in Figure 4.36.

```
1 #include "clang/Frontend/FrontendPluginRegistry.h"
2
3 #include "ClassAction.hpp"
4
5 static
  ↪ clang::FrontendPluginRegistry::Add<clangbook::classchecker::ClassAction>
6   X("classchecker", "Checks the complexity of C++ classes");
```

Figure 4.36: Source code for ClassChecker.cpp

As we can observe, the majority of the initialization is hidden by helper classes, and we only need to pass our implementation to `lang::FrontendPluginRegistry::Add`.

Now we are ready to build and test our clang plugin.

4.6.7 Building and running plugin code

We need to specify a path to the installation folder for our LLVM project. The rest of the procedure is the standard one that we have previously used, see Figure 3.11:

```
export LLVM_HOME=<...>/llvm-project/install
mkdir build
cd build
cmake -G Ninja -DCMAKE_BUILD_TYPE=Debug ..
ninja classchecker
```

Figure 4.37: Configure and build commands for the Clang plugin

The build artifacts will be located in the `build` folder. We can then run our plugin on a test file as follows, where `<filepath>` is the file we want to compile:

```
$ <...>/llvm-project/install/bin/clang -fsyntax-only   \
             -fplugin=./build/libclasschecker.so   \
             <filepath>
```

Figure 4.38: How to run the Clang plugin on a test file

For example, if we use a test file named `test.cpp` that defines a class with three methods (see Figure 4.39), we will not receive any warnings.

```
1 class Simple {
2 public:
3   void func1() {}
4   void func2() {}
5   void func3() {}
6 };
```

Figure 4.39: Test for the clang plugin: test.cpp

However, if we specify a smaller threshold, we will receive a warning for the file:

```
$ <...>/llvm-project/install/bin/clang -fsyntax-only    \
             -fplugin-arg-classchecker-threshold=2 \
             -fplugin=./build/libclasschecker.so    \
             test.cpp
test.cpp:1:7: warning: class Simple is too complex: method count = 3
    1 | class Simple {
      |       ^
1 warning generated.
```

Figure 4.40: Clang plugin run on test.cpp

It's now time to create a LIT test for our plugin.

4.6.8 LIT tests for clang plugin

We'll begin with a description of the project organization. We'll adopt the common pattern used in the clang source code and place our tests in the test folder. This folder will contain the following files:

- lit.site.cfg.py.in : This is the main configuration file, a CMake config file. It replaces patterns marked as '@...@' with corresponding values defined during the CMake configuration. Additionally, this file loads lit.cfg.py .

- lit.cfg.py : This serves as the primary configuration file for LIT tests.

- simple_test.cpp : This is our LIT test file.

The basic workflow is as follows: CMake takes lit.site.cfg.py.in as a template and generates the corresponding lit.site.cfg.py in the build/test folder. This file is then utilized by LIT tests as a seed to execute the tests.

LIT config files

There are two configuration files for LIT tests. The first one is shown in Figure 4.41.

```
1 config.ClassComplexityChecker_obj_root = "@CMAKE_CURRENT_BINARY_DIR@"
2 config.ClassComplexityChecker_src_root = "@CMAKE_CURRENT_SOURCE_DIR@"
3 config.ClangBinary = "@LLVM_HOME@/bin/clang"
4 config.FileCheck = "@FILECHECK_COMMAND@"
5
6 lit_config.load_config(
7         config, os.path.join(config.ClassComplexityChecker_src_root,
          ↪ "test/lit.cfg.py"))
```

Figure 4.41: lit.site.cfg.py.in file

This file is a CMake template that will be converted into a Python script. The most crucial part is shown in *Lines 6-7*, where the main LIT config is loaded. It is sourced from the main source tree and is not copied to the build folder.

The subsequent configuration is displayed in Figure 4.42. It is a Python script containing the primary configuration for LIT tests.

```python
1  # lit.cfg.py
2  import lit.formats
3
4  config.name = 'classchecker'
5  config.test_format = lit.formats.ShTest(True)
6  config.suffixes = ['.cpp']
7  config.test_source_root = os.path.dirname(__file__)
8
9  config.substitutions.append(('%clang-binary', config.ClangBinary))
10 config.substitutions.append(('%path-to-plugin',
       ↪  os.path.join(config.ClassComplexityChecker_obj_root,
       ↪  'libclasschecker.so')))
11 config.substitutions.append(('%file-check-binary', config.FileCheck))
```

Figure 4.42: lit.cfg.py file

Lines 4-7 define the fundamental configuration; for example, *Line 6* determines which files should be utilized for tests. All files with the '.cpp' extension in the test folder will be employed as LIT tests.

Lines 9-11 detail the substitutions that will be employed in the LIT tests. These include the path to the clang binary (*Line 9*), the path to the shared library with the plugin (*Line 10*), and the path to the FileCheck utility (*Line 11*).

We have defined only one basic LIT test, simple_test.cpp , as shown in Figure 4.43.

```
 1 // RUN: %clang-binary -fplugin=%path-to-plugin -fsyntax-only %s 2>&1 |
   ↪ %file-check-binary %s
 2
 3 class Simple {
 4 public:
 5   void func1() {}
 6   void func2() {}
 7 };
 8
 9 // CHECK: :[[@LINE+1]]:{{[0-9]+}}: warning: class Complex is too complex:
   ↪ method count = 6
10 class Complex {
11 public:
12   void func1() {}
13   void func2() {}
14   void func3() {}
15   void func4() {}
16   void func5() {}
17   void func6() {}
18 };
```

Figure 4.43: simple_test.cpp file

The use of substitutions can be observed in *Line 1*, where paths to the clang binary, the plugin shared library, and the FileCheck utility are referenced. Special patterns recognized by the utility are used in *Line 9*.

The final piece of the puzzle is the CMake configuration. This will set up the required variables for substitutions in lit.site.cfg.py.in and also define a custom target to run the LIT tests.

CMake configuration for LIT tests

The CMakeLists.txt file requires some adjustments to support LIT tests. The necessary changes are displayed in Figure 4.44.

```
31   find_program(LIT_COMMAND llvm-lit PATH ${LLVM_BUILD}/bin)

32   find_program(FILECHECK_COMMAND FileCheck ${LLVM_BUILD}/bin)

33   if(LIT_COMMAND AND FILECHECK_COMMAND)

34     message(STATUS "$LIT_COMMAND found: ${LIT_COMMAND}")

35     message(STATUS "$FILECHECK_COMMAND found: ${FILECHECK_COMMAND}")

36

37     # Point to our custom lit.cfg.py

38     set(LIT_CONFIG_FILE "${CMAKE_CURRENT_SOURCE_DIR}/test/lit.cfg.py")

39

40     # Configure lit.site.cfg.py using current settings

41     configure_file("${CMAKE_CURRENT_SOURCE_DIR}/test/lit.site.cfg.py.in"

42                    "${CMAKE_CURRENT_BINARY_DIR}/test/lit.site.cfg.py"

43                    @ONLY)

44

45     # Add a custom target to run tests with lit

46     add_custom_target(check-classchecker

47                       COMMAND ${LIT_COMMAND} -v
                          ↪  ${CMAKE_CURRENT_BINARY_DIR}/test

48                       COMMENT "Running lit tests for classchecker clang
                          ↪  plugin"

49                       USES_TERMINAL)

50   else()

51     message(FATAL_ERROR "It was not possible to find the LIT executables at
         ↪  ${LLVM_BUILD}/bin")

52   endif()
```

Figure 4.44: LIT tests configuration at CMakeLists.txt

In *Lines 31 and 32*, we search for the necessary utilities, llvm-lit and FileCheck . It's worth noting that they rely on the $LLVM_BUILD environment variable, which we also verify in *Line 12* of the config (see Figure 4.32). The steps in *Lines 41-43* are essential for generating lit.site.cfg.py from the provided template file, lit.site.cfg.py.in. Lastly, we establish a custom target to execute the LIT tests in *Lines 46-49*.

Now we are ready to start the LIT tests.

Running LIT tests

To initiate the LIT tests, we must set an environment variable that points to the build folder, compile the project, and then execute the custom target, check-classchecker. Here's how this can be done:

```
export LLVM_BUILD=<...>/llvm-project/build
export LLVM_HOME=<...>/llvm-project/install
rm -rf build; mkdir build; cd build
cmake -G Ninja -DCMAKE_BUILD_TYPE=Debug ..
ninja classchecker
ninja check-classchecker
```

Figure 4.45: Configure, build and check commands for the Clang plugin

Upon executing these commands, you may observe the following output:

```
...
[2/2] Linking CXX shared module libclasschecker.so
[0/1] Running lit tests for classchecker clang plugin
-- Testing: 1 tests, 1 workers --
PASS: classchecker :: simple_test.cpp (1 of 1)

Testing Time: 0.12s
Passed: 1
```

Figure 4.46: LIT test execution

With this, we conclude our first comprehensive project, which encompasses a practical clang plugin that can be tailored via supplemental plugin arguments. Additionally, it includes the respective tests that can be executed to verify its functionality.

4.7 Summary

In this chapter, we became familiar with the basic classes from the LLVM ADT library. We gained knowledge of Clang diagnostics and the test frameworks used in LLVM for various types of testing. Using this knowledge, we created a simple Clang plugin that detects complex classes and issues a warning about their complexity.

The chapter concludes the first part of the book, where we gained basic knowledge of the Clang compiler frontend. We are now prepared to explore various tools built on the foundation of Clang libraries. We will begin with Clang-Tidy, a powerful linter framework used to detect various issues in C++ code.

4.8 Further reading

- LLVM Coding Standards: `https://llvm.org/docs/CodingStandards.html`

- LLVM Programmer's Manual: `https://llvm.org/docs/ProgrammersManual.html`

- "Clang" CFE Internals Manual: `https://clang.llvm.org/docs/InternalsManual.html`

- How to set up LLVM-style RTTI for your class hierarchy: `https://llvm.org/docs/HowToSetUpLLVMStyleRTTI.html`

- LIT - LLVM Integrated Tester: `https://llvm.org/docs/CommandGuide/lit.html`

Part 2
Clang Tools

You can find some info about different Clang tools here. We will start with linters that are based on Clang-Tidy, continue with some advanced code analysis techniques (CFG and live time analysis). The next chapter will be about different refactoring tools such as Clang-Format. The last chapter will be about IDE support. We are going to investigate how Visual Studio Code can be extended with language server provided by LLVM (Clangd).

This part has the following chapters:

- *Chapter 5, Clang-Tidy Linter Framework*

- *Chapter 6, Advanced Code Analysis*

- *Chapter 7, Refactoring Tools*

- *Chapter 8, IDE Support and Clangd*

5

Clang-Tidy Linter Framework

This chapter introduces Clang-Tidy, the clang-based linter framework that utilizes the **Abstract Syntax Tree (AST)** to identify anti-patterns in C/C++/Objective-C code. First, we'll discuss Clang-Tidy's capabilities, the types of checks it offers, and how to use them. After that, we will delve into the architecture of Clang-Tidy and explore how to create our own custom lint check. In this chapter, we'll cover the following topics:

- An overview of Clang-Tidy, including a brief description of the different checks provided by default

- The internal design of Clang-Tidy

- How to create a custom Clang-Tidy check

5.1   Technical requirements

The source code for this chapter is located in the `chapter5` folder of the book's GitHub repository: `https://github.com/PacktPublishing/Clang-Compiler-Frontend-Packt /tree/main/chapter5`.

5.2   Overview of Clang-Tidy and usage examples

Clang-Tidy is a linter and static analysis tool for C and C++ code. It is a part of the Clang and LLVM project. The tool is built on top of the Clang frontend, which means it understands your code in depth, giving it the ability to catch a wide range of issues.

Here are some key points to understand about Clang-Tidy:

- **Checks**: Clang-Tidy contains a series of "checks" that identify various issues or suggest enhancements. These checks range from performance improvements and potential bugs to coding style and modern C++ best practices. For instance, it might suggest using `emplace_back` instead of `push_back` for certain cases or identify areas where you might be accidentally using integer overflow.

- **Extensibility**: New checks can be added to Clang-Tidy, making it a highly extensible tool. If you have specific coding guidelines or practices you want to enforce, you can write a check for it.

- **Integration**: Clang-Tidy is often used within CI/CD pipelines or integrated with development environments. Many IDEs support Clang-Tidy directly or via plugins, so you can get real-time feedback on your code as you write it.

- **Automatic fixes**: One of the powerful features of Clang-Tidy is its ability to not only identify issues but also automatically fix many of them. This is done with the `-fix` option. It is, however, important to review the proposed changes, as automatic fixes might not always be perfect.

- **Configuration**: You can configure which checks Clang-Tidy performs using a configuration file or command-line options. This allows teams to enforce specific coding standards or prioritize certain types of issues. For example, the `-checks='-*,modernize-*'` command-line option will disable all checks but not the checks from modernize set.

- **Modern C++ best practices**: One of the often-appreciated features of Clang-Tidy is its emphasis on modern C++ idioms and best practices. It can guide developers to write safer, more performant, and more idiomatic C++ code.

After acquiring basic knowledge about Clang-Tidy, let's examine how it can be built.

5.2.1 Building and testing Clang-Tidy

We will use the basic build configuration specified in Figure 1.4 and build Clang-Tidy with the following Ninja command:

```
$ ninja clang-tidy
```

Figure 5.1: Using the Ninja command to build Clang-Tidy

We can install the Clang-Tidy binary to the designated `install` folder using the following command:

```
$ ninja install-clang-tidy
```

Figure 5.2: Using the Ninja command to install Clang-Tidy

Using the build configuration from Figure 1.4, the command will install the Clang-Tidy binary under the `<...>/llvm-project/install/bin` folder. Here, `<...>/llvm-project` refers to the path where the LLVM code base was cloned (see Figure 1.1).

> **Important note**
>
> If you use a build configuration with shared libraries (with the `BUILD_SHARED_LIBS` flag set to `ON`), as shown in Figure 1.12, then you might need to install and built all artifacts with `ninja install`.

Clang-Tidy is part of Clang-Tools-Extra, and its tests are a part of the `clang-tools` CMake target. Thus, we can run the tests with the following command:

```
$ ninja check-clang-tools
```

Figure 5.3: Using the Ninja command to run Clang-Tidy tests

The command will run LIT tests (see *Section 4.5.2, LLVM test framework*) for all Clang-Tidy checks, and will also run unit tests for the Clang-Tidy core system. You can also run a specific LIT test separately; for example, if we want to run the LIT test for the `modernize-loop-convert` check, we can use the following command:

```
$ cd <...>/llvm-project
$ build/bin/llvm-lit -v \
clang-tools-extra/test/clang-tidy/checkers/modernize/loop-convert-basic.cpp
```

Figure 5.4: Testing the modernize-loop-convert clang-tidy check

The command will produce the following output:

```
-- Testing: 1 tests, 1 workers --
PASS: Clang Tools :: clang-tidy/checkers/modernize/loop-convert-basic.cpp
(1 of 1)

Testing Time: 1.38s
  Passed: 1
```

Figure 5.5: LIT test output for the cppcoreguidelines-owning-memory clang-tidy check

After building and testing Clang-Tidy, it's now time to run it on some code examples.

5.2.2 Clang-Tidy usage

To test Clang-Tidy, we will use the following test program:

```cpp
1 #include <iostream>
2 #include <vector>
3
4 int main() {
5     std::vector<int> numbers = {1, 2, 3, 4, 5};
6     for (std::vector<int>::iterator it = numbers.begin(); it !=
         ↪ numbers.end();
7             ++it) {
8         std::cout << *it << std::endl;
9     }
10    return 0;
11 }
```

Figure 5.6: Test program for Clang-Tidy: loop-convert.cpp

The program is correctly written in the older C++ code style, that is, before C++11. Clang-Tidy has a set of checks that encourage adopting the modern C++ code style and using new C++ idioms available in the latest C++ standard. These checks can be run on the program as follows:

```
1 $ <...>/llvm-project/install/bin/clang-tidy \
2   -checks='-*,modernize-*'                    \
3   loop-convert.cpp                            \
4   -- -std=c++17
```

Figure 5.7: Running Clang-Tidy modernize checks on loop-convert.cpp

The most important parts of Figure 5.7 are as follows:

- *Line 1*: The path to the Clang-Tidy binary is specified here.

- *Line 2*: We remove all checks using the '-* ' option. Then, we enable all checks with the 'modernize ' prefix by using the '-*,modernize-* ' value for the '-checks ' argument.

- *Line 3*: We specify the path to the code to be tested.

- *Line 4*: We pass additional arguments to the compiler, notably specifying that we want the compiler to use C++17 as the C++ standard.

The output of the program will be as follows:

```
loop-convert.cpp:4:5: warning: use a trailing return type for this function
...
    4 | int main() {
      | ~~~ ^
      | auto        -> int
loop-convert.cpp:6:3: warning: use range-based for loop instead
[modernize-loop-convert]
    6 |    for (std::vector<int>::iterator it = numbers.begin();
           it != numbers.end();
      |     ^   ~~~~~~~~~~~~~~~~~~~~~~~~~~~
      |          ~~~~~~~~~~~~~~~~~~~~~~~~~~~~~~~~~~~~~~~~~
      |        (int & number : numbers)
    7 |        ++it) {
      |        ~~~~~
    8 |     std::cout << *it << std::endl;
      |                 ~~~
      |                 number
loop-convert.cpp:6:8: warning: use auto when declaring iterators
[modernize-use-auto]
    6 |    for (std::vector<int>::iterator it = numbers.begin();
           it != numbers.end();
```

```
    |        ^
```

note: this fix will not be applied because it overlaps with another fix

Figure 5.8: Output from running Clang-Tidy on loop-convert.cpp

As we can see, several issues were detected, and Clang-Tidy suggested some fixes. Unfortunately, some of them conflict with each other, especially modernize-loop-convert and modernize-use-auto , and cannot be applied together. On the other hand, we can apply the fix suggested by modernize-loop-convert by running only this specific check to avoid any conflicts, as follows:

```
1  $ <...>/llvm-project/install/bin/clang-tidy \
2    -checks='-*,modernize-loop-convert'        \
3    -fix                                       \
4    loop-convert.cpp                           \
5    -- -std=c++17
```

Figure 5.9: Running a modernize-loop-convert check on loop-convert.cpp

As we can see, the second line has changed compared to Figure 5.7, and another line (3) has been added. The latter instructs Clang-Tidy to apply the fixes suggested by the check. The resulting code can be found in the original file:

```
1 #include <iostream>
2 #include <vector>
3
4 int main() {
5   std::vector<int> numbers = {1, 2, 3, 4, 5};
6   for (int & number : numbers) {
7     std::cout << number << std::endl;
8   }
9   return 0;
10 }
```

Figure 5.10: Fixed test program for Clang-Tidy: loop-convert.cpp

As we can see, *Lines 6* and *7* have changed compared to the original code from Figure 5.6. This functionality makes Clang-Tidy a powerful tool that can not only detect issues but also fix them. We will explore this possibility in greater depth later in *Section 7.3, Clang-Tidy as a code modification tool.*

5.2.3 Clang-Tidy checks

Clang-Tidy has a wide variety of checks grouped into different categories. Here's a concise list of some of the main categories, with an example check from each and a brief description:

1. **boost-***:

   - boost-use-to-string: Suggests replacing boost::lexical_cast<std::string> with boost::to_string

2. **bugprone-***:

   - bugprone-integer-division: Warns when integer division in a floating-point context is likely to cause unintended loss of precision

3. **cert-*** (Checks related to the CERT C++ Secure Coding Standard):

   - cert-dcl03-c: Ensures that macros are not used in unsafe contexts

4. **cppcoreguidelines-*** (Checks from C++ Core Guidelines):

   - `cppcoreguidelines-slicing`: Warns on slicing (object slicing, where a derived object is assigned to a base object, cutting off the derived parts)

5. **google-*** (Google's coding conventions):

   - `google-build-using-namespace`: Flags using-directives

6. **llvm-*** (LLVM coding conventions):

   - `llvm-namespace-comment`: Ensures that namespaces have closing comments

7. **misc-*** (Miscellaneous checks):

   - `misc-unused-parameters`: Flags parameters that are unused

8. **modernize-*** (Modernization checks for C++):

   - `modernize-use-auto`: Recommends the use of `auto` for variable declarations when appropriate

9. **performance-***:

   - `performance-faster-string-find`: Suggests faster alternatives for string searching

10. **readability-***:

    - `readability-identifier-naming`: Ensures consistent identifier naming

This list is just a representation of a subset of the checks available. Each category contains multiple checks, and there are additional categories in the tool as well. For a complete, up-to-date list of checks and their detailed descriptions, refer to the official Clang-Tidy documentation [17] or use the `clang-tidy -list-checks` command on your system.

After learning how to build and use clang-tidy, it's time to delve deeper and examine its internal design.

5.3   Clang-Tidy's internal design

Clang-Tidy is built on top of Clang. At its core, Clang-Tidy leverages Clang's ability to parse and analyze source code into an AST. Each check in Clang-Tidy essentially involves defining patterns or conditions to match against this AST. When a match is found, a diagnostic can be raised, and in many cases, an automatic fix can be suggested. The tool operates on the basis of individual "checks" that target specific issues or coding styles. Checks are implemented as plugins, making Clang-Tidy extensible. The `ASTMatchers` library facilitates writing these checks by providing a domain-specific language to query the AST; see *Section 3.5, AST matchers* and the official documentation [16] for more info. This ensures that checks are both concise and expressive. Clang-Tidy also has support for analyzing the code base using a compilation database, which provides context such as compile flags (see *Chapter 9, Appendix 1: Compilation Database* for more info). This comprehensive integration with Clang's internals makes Clang-Tidy a powerful static analysis tool with precise code transformation capabilities.

5.3.1   Internal organization

The internal organization of clang-tidy within the Clang code base can be complex due to its deep integration with the Clang libraries, but at a high level, the organization can be broken down as follows:

1. **Source and headers**: The main source code and headers for `clang-tidy` are located in the `clang-tools-extra` repository, specifically within the `clang-tidy` directory.

2. **Main driver**: The `ClangTidyMain.cpp` file, located in the `tool` subfolder, serves as the main driver for the Clang-Tidy tool.

3. **Core infrastructure**: Files such as `ClangTidy.cpp`, `ClangTidy.h` manage the core functionalities and options.

4. **Checks**: Checks are organized into subdirectories based on categories (e.g., `bugprone` or `modernize`).

5. **Utilities**: The `utils` directory contains utility classes and functions.

6. **AST Matchers**: The ASTMatchers library, which we explored previously in *Section 3.5, AST matchers*, is integral for querying the AST.

7. **Clang diagnostics**: Clang-Tidy actively uses the Clang diagnostics subsystem to print diagnostics messages and suggest fixes (see *Section 4.4.2, Diagnostics support*).

8. **Tests**: Tests are located in the test directory and use LLVM's LIT framework (see *Section 4.5.2, LLVM test framework*). It's worth noting that the test folder is shared with other projects inside the clang-tools-extra folder.

9. **Documentation**: The docs directory contains documentation for Clang-Tidy. As well as the tests, the documentation is a part of other projects inside the
clang-tools-extra folder.

These relationships are schematically illustrated in the following figure:

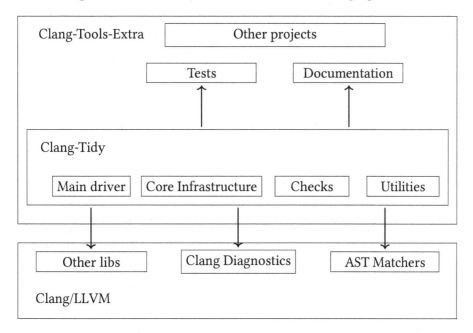

Figure 5.11: Clang-Tidy's internal organization

Now that we have gained an understanding of Clang-Tidy's internals and its relationship with other parts of Clang/LLVM, it's time to explore components external to the Clang-Tidy binary: its configuration and other tools that leverage the functionality provided by Clang-Tidy.

5.3.2   Configuration and integration

The Clang-Tidy binary can interact with other components, as shown in Figure 5.12.

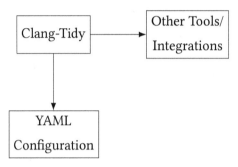

Figure 5.12: Clang-Tidy external components: configuration and integrations

Clang-Tidy can be seamlessly integrated with various **Integrated Development Environments (IDEs)**, such as Visual Studio Code, CLion, and Eclipse, to provide real-time feedback during coding. We will explore this possibility later in *Section 8.5.2, Clang-Tidy*.

It can also be incorporated into build systems such as CMake and Bazel to run checks during builds. **Continuous Integration (CI)** platforms, such as Jenkins and GitHub Actions, often employ Clang-Tidy to ensure code quality on pull requests. Code review platforms, such as Phabricator, utilize Clang-Tidy for automated reviews. Additionally, custom scripts and static analysis platforms can harness Clang-Tidy's capabilities for tailored workflows and combined analyses.

Another important part of Clang-Tidy shown in Figure 5.12 is its configuration. Let's explore it in detail.

Clang-Tidy configuration

Clang-Tidy uses a configuration file to specify which checks to run and to set options for those checks. This configuration is done using a `.clang-tidy` file.

The .clang-tidy file is written in YAML format. It typically contains two main keys: Checks and CheckOptions .

We will begin with the Checks key, which allows us to specify which checks to enable or disable:

- Use - to disable a check

- Use * as a wildcard to match multiple checks

- Checks are comma-separated

Here's an example:

```
1    Checks: '-*,modernize-*'
```

Figure 5.13: Checks key of a .clang-tidy config file

The next key is CheckOptions . This key allows us to set options for specific checks, with each option specified as a key-value pair. An example is provided here:

```
1 CheckOptions:
2    - key: readability-identifier-naming.NamespaceCase
3      value: CamelCase
4    - key: readability-identifier-naming.ClassCase
5      value: CamelCase
```

Figure 5.14: CheckOptions key of a .clang-tidy config file

When Clang-Tidy is run, it searches for the .clang-tidy file in the directory of the file being processed and its parent directories. The search stops when the file is found.

Now that we have an understanding of Clang-Tidy's internal design, it's time to create our first custom Clang-Tidy check using the information we've gathered from this and previous chapters of the book.

5.4   Custom Clang-Tidy check

In this part of the chapter, we will transform our plugin example (see *Section 4.6, Clang plugin project*) into a Clang-Tidy check. This check will estimate the complexity of a C++ class based on the number of methods it contains. We will define a threshold as a parameter for the check.

Clang-Tidy offers a tool designed to aid in the creation of checks. Let's begin by creating a skeleton for our check.

5.4.1   Creating a skeleton for the check

Clang-Tidy provides a specific Python script, add_new_check.py , to assist in creating new checks. This script is located in the clang-tools-extra/clang-tidy directory. The script requires two positional parameters:

- module : This refers to the module directory where the new tidy check will be placed. In our case, this will be misc .

- check : This is the name of the new tidy check to add. For our purposes, we will name it classchecker .

By running the script in the llvm-project directory (which contains the cloned LLVM repository), we receive the following output:

```
$ ./clang-tools-extra/clang-tidy/add_new_check.py misc classchecker
...
Updating ./clang-tools-extra/clang-tidy/misc/CMakeLists.txt...
Creating ./clang-tools-extra/clang-tidy/misc/ClasscheckerCheck.h...
Creating ./clang-tools-extra/clang-tidy/misc/ClasscheckerCheck.cpp...
Updating ./clang-tools-extra/clang-tidy/misc/MiscTidyModule.cpp...
Updating clang-tools-extra/docs/ReleaseNotes.rst...
Creating clang-tools-extra/test/clang-tidy/checkers/misc/classchecker.cpp...
Creating clang-tools-extra/docs/clang-tidy/checks/misc/classchecker.rst...
Updating clang-tools-extra/docs/clang-tidy/checks/list.rst...
```

```
Done. Now it's your turn!
```

Figure 5.15: Creating a skeleton for the misc-classchecker check

From the output, we can observe that several files under the `clang-tools-extra/` `clang-tidy` directory have been updated. These files pertain to checks registration, such as `misc/MiscTidyModule.cpp`, or build configuration, such as `misc/CMakeLists.txt`. The script also generated several new files, which we need to modify in order to implement our check's desired logic:

- `misc/ClasscheckerCheck.h` : This is the header file for our check

- `misc/ClasscheckerCheck.cpp` : This file will house the implementation of our check

Additionally, the script has generated a LIT test for our check, named `ClassChecker.cpp`. This test can be found in the `clang-tools-extra/test/clang-tidy/checkers/misc` directory.

Apart from the source files, the script also modifies some documentation files in the `clang-tools-extra/docs` directory:

- `ReleaseNotes.rst` : This file contains updated release notes with placeholder entries for our new check

- `clang-tidy/checks/misc/classchecker.rst` : This serves as the primary documentation for our check

- `clang-tidy/checks/list.rst` : The list of checks has been updated to include our new check alongside other checks from the 'misc ' module

We will now turn our attention to implementing the check and the subsequent build process.

5.4.2 Clang-Tidy check implementation

We'll begin by modifying `ClasscheckerCheck.cpp`. The generated file can be found in the `clang-tools-extra/clang-tidy/misc` directory. Let's replace the generated code with the following (note: the generated comment containing the license info has been omitted for brevity):

```cpp
 9 #include "ClasscheckerCheck.h"
10 #include "clang/AST/ASTContext.h"
11 #include "clang/ASTMatchers/ASTMatchFinder.h"
12 using namespace clang::ast_matchers;
13
14 namespace clang::tidy::misc {
15 void ClasscheckerCheck::registerMatchers(MatchFinder *Finder) {
16     // Match every C++ class.
17     Finder->addMatcher(cxxRecordDecl().bind("class"), this);
18 }
19 void ClasscheckerCheck::check(const MatchFinder::MatchResult &Result) {
20     const auto *ClassDecl = Result.Nodes.getNodeAs<CXXRecordDecl>("class");
21     if (!ClassDecl || !ClassDecl->isThisDeclarationADefinition())
22         return;
23     unsigned MethodCount = 0;
24     for (const auto *D : ClassDecl->decls()) {
25         if (isa<CXXMethodDecl>(D))
26             MethodCount++;
27     }
28     unsigned Threshold = Options.get("Threshold", 5);
29     if (MethodCount > Threshold) {
30         diag(ClassDecl->getLocation(),
31             "class %0 is too complex: method count = %1",
32             DiagnosticIDs::Warning)
33         << ClassDecl->getName() << MethodCount;
34     }
35 }
36 } // namespace clang::tidy::misc
```

Figure 5.16: Modifications to ClasscheckerCheck.cpp

We replaced the original stub with *Lines 15-35* to implement the necessary changes.

To integrate our check into the Clang-Tidy binary, we can execute the standard build procedure from the `build` directory within the LLVM source tree; see Figure 5.2.

The name of our check is defined in the modified `MiscTidyModule.cpp` file in `clang-tools-extra/clang-tidy/misc` folder:

```
40 class MiscModule : public ClangTidyModule {
41 public:
42   void addCheckFactories(ClangTidyCheckFactories &CheckFactories) override
     ↪ {
43     CheckFactories.registerCheck<ClasscheckerCheck>(
44         "misc-classchecker");
45     CheckFactories.registerCheck<ConfusableIdentifierCheck>(
46         "misc-confusable-identifiers");
```

Figure 5.17: Modifications to MiscTidyModule.cpp

As illustrated in Figure 5.17 (*Lines 43-44*), we registered the new check under the name `"misc-classchecker"`. After the code modification, we are ready to recompile Clang-Tidy with

```
$ ninja install
```

We can verify that the check has been added by executing Clang-Tidy with the `-list-checks` argument as follows:

```
<...>/llvm-project/install/bin/clang-tidy -checks '*' -list-checks
...
    misc-classchecker
...
```

Figure 5.18: Clang-Tidy `-list-checks` option

It's worth noting that we enabled all checks using the `-checks '*'` command-line option, as shown in Figure 5.18.

To test the check, we can use the file from the clang plugin project, as seen in Figure 4.39:

```
1 class Simple {
2 public:
3   void func1() {}
4   void func2() {}
5   void func3() {}
6 };
```

Figure 5.19: Test file for the misc-classchecker clang-tidy check: test.cpp

This file contains three methods. To trigger a warning, we must set the threshold to 2, as demonstrated:

```
1 $ <...>/llvm-project/install/bin/clang-tidy                        \
2   -checks='-*,misc-classchecker'                                    \
3   -config="{CheckOptions: [{key:misc-classchecker.Threshold, value:'2'}]}"\
4   test.cpp                                                          \
5   -- -std=c++17
```

Figure 5.20: Run a misc-classchecker check on the test file: test.cpp

The output will be as follows:

```
test.cpp:1:7: warning: class Simple is too complex: method count = 3
[misc-classchecker]
class Simple {
      ^
```

Figure 5.21: Output of the misc-classchecker check for the test.cpp test file

After testing the file with custom source code, it's time to create an LIT test for our check.

5.4.3 LIT test

For the LIT test, we will use the slightly modified code from Figure 4.43. Let's modify classchecker.cpp, located in the clang-tools-extra/test/clang-tidy/checkers/misc folder, as follows:

```
1  // RUN: %check_clang_tidy %s misc-classchecker %t
2
3  class Simple {
4  public:
5    void func1() {}
6    void func2() {}
7  };
8
9  // CHECK-MESSAGES: :[[@LINE+1]]:{{[0-9]+}}: warning: class Complex is too
   ↪  complex: method count = 6 [misc-classchecker]
10 class Complex {
11 public:
12   void func1() {}
13   void func2() {}
14   void func3() {}
15   void func4() {}
16   void func5() {}
17   void func6() {}
18 };
```

Figure 5.22: LIT test: classchecker.cpp

As we can see, the only difference compared to Figure 4.43 is in *Line 1*, where we specify which commands should be run, and in *Line 9*, where we define the test pattern.

We can run the test as follows:

```
$ cd <...>/llvm-project
$ build/bin/llvm-lit -v \
  clang-tools-extra/test/clang-tidy/checkers/misc/classchecker.cpp
```

Figure 5.23: Testing the misc-classchecker clang-tidy check

The command produces the following output:

```
-- Testing: 1 tests, 1 workers --
PASS: Clang Tools :: clang-tidy/checkers/misc/classchecker.cpp (1 of 1)

Testing Time: 0.12s
  Passed: 1
```

Figure 5.24: Testing output for misc-classchecker

We can also use the command shown in Figure 5.3 to run all clang-tidy checks, including our newly added one.

When we run our check on a real code base, as opposed to synthetic tests, we may encounter unexpected results. One such issue has already been discussed in *Section 3.7, Processing AST in the case of errors* and pertains to the impact of compilation errors on Clang-Tidy results. Let's delve into this problem using a specific example.

5.4.4   Results in the case of compilation errors

Clang-Tidy uses AST as the information provider for checks, and the checks can produce wrong results if the information source is broken. The typical case is when the analyzed code has compilation errors (see *Section 3.7, Processing AST in the case of errors*).

Consider the following code as an example:

```
1 class MyClass {
2 public:
3   void doSomething();
4 };
5
6 void MyClass::doSometing() {}
```

Figure 5.25: Test file with compilation errors: error.cpp

In the example, we made a syntax error in *Line 6*: the method name is incorrectly written as 'doSometing' instead of 'doSome**th**ing'. If we run our check on the file without any parameters, we will receive the following output:

```
error.cpp:1:7: warning: class MyClass is too complex: method count = 7
[misc-classchecker]
class MyClass {
      ^
error.cpp:6:15: error: out-of-line definition of 'doSometing' ...
[clang-diagnostic-error]
void MyClass::doSometing() {}
              ^~~~~~~~~~~
              doSomething
error.cpp:3:8: note: 'doSomething' declared here
  void doSomething();
       ^
Found compiler error(s).
```

Figure 5.26: Running a misc-classchecker check on a file containing compilation errors

Our check seems to be working incorrectly with this code. It assumes the class has seven methods when, in fact, it has only one.

The case of compilation errors can be considered an edge case, and we can process it correctly. Before addressing these cases, we should investigate the produced AST to examine the issue.

5.4.5   Compilation errors as edge cases

Let's use `clang-query` (see *Section 3.6, Explore Clang AST with clang-query*) to explore what has happened with the AST. The program with the error fixed is shown in the following figure:

```cpp
1 class MyClass {
2 public:
3   void doSomething();
4 };
5
6 void MyClass::doSomething() {}
```

Figure 5.27: noerror.cpp test file with compilation errors fixed

The `clang-query` command can be run on the file as follows:

```
$ <...>/llvm-project/install/bin/clang-query noerror.cpp -- --std=c++17
```

Figure 5.28: Clang-Query run on noerror.cpp file with compilation errors fixed

Then, we will set up Clang-Query's output as dump and find all matches for `CXXRecordDecl`

```
clang-query> set output dump
clang-query> match cxxRecordDecl()
```

Figure 5.29: Setup Clang-Query output and run matchers

The result is shown below

```
Match #1:


Binding for "root":
CXXRecordDecl ... <noerror.cpp:1:1, line:4:1> line:1:7 class MyClass
definition
|-DefinitionData ...
| |-DefaultConstructor exists trivial ...
| |-CopyConstructor simple trivial ...
| |-MoveConstructor exists simple trivial ...
| |-CopyAssignment simple trivial ...
| |-MoveAssignment exists simple trivial ...
| `-Destructor simple irrelevant trivial ...
|-CXXRecordDecl ... <col:1, col:7> col:7 implicit class MyClass
|-AccessSpecDecl ... <line:2:1, col:7> col:1 public
`-CXXMethodDecl ... <line:3:3, col:20> col:8 doSomething 'void ()'
...
```

Figure 5.30: AST for the noerror.cpp file with compilation errors fixed

Compare it with the output for the code with an error (see Figure 5.25). We run Clang-Query on the error.cpp file and set up the required matcher as follows

```
$ <...>/llvm-project/install/bin/clang-query error.cpp -- --std=c++17
clang-query> set output dump
clang-query> match cxxRecordDecl()
```

Figure 5.31: Clang-Query run on error.cpp

The found match is shown below:

```
CXXRecordDecl ... <error.cpp:1:1, line:4:1> line:1:7 class MyClass
definition
|-DefinitionData ...
| |-DefaultConstructor exists trivial ...
| |-CopyConstructor simple trivial ..
| |-MoveConstructor exists simple trivial
| |-CopyAssignment simple trivial ...
| |-MoveAssignment exists simple trivial
| `-Destructor simple irrelevant trivial
|-CXXRecordDecl ... <col:1, col:7> col:7 implicit class MyClass
|-AccessSpecDecl ... <line:2:1, col:7> col:1 public
|-CXXMethodDecl ... <line:3:3, col:20> col:8 doSomething 'void ()'
|-CXXConstructorDecl ... <line:1:7> col:7 implicit constexpr MyClass
'void ()' ...
|-CXXConstructorDecl ... <col:7> col:7 implicit constexpr MyClass
'void (const MyClass &)' ...
| `-ParmVarDecl ... <col:7> col:7 'const MyClass &'
|-CXXMethodDecl ... <col:7> col:7 implicit constexpr operator= 'MyClass
&(const MyClass &)' inline default trivial ...
| `-ParmVarDecl ... <col:7> col:7 'const MyClass &'
|-CXXConstructorDecl ... <col:7> col:7 implicit constexpr MyClass 'void
(MyClass &&)' ...
| `-ParmVarDecl ... <col:7> col:7 'MyClass &&'
|-CXXMethodDecl ... <col:7> col:7 implicit constexpr operator= 'MyClass
&(MyClass &&)' ...
| `-ParmVarDecl ... <col:7> col:7 'MyClass &&'
`-CXXDestructorDecl ... <col:7> col:7 implicit ~MyClass 'void ()' inline
default ...
...
```

Figure 5.32: AST for the error.cpp file with a compilation error

As we can see, all additional methods are added implicitly. We can exclude them by modifying *Line 30* (see Figure 5.16) of our check code, as shown:

```
29      for (const auto *D : ClassDecl->decls()) {
30          if (isa<CXXMethodDecl>(D) && !D->isImplicit())
31              MethodCount++;
32      }
```

Figure 5.33: Exclude implicit declaration from the check report

If we run the modified check on the file that contains compilation errors, we will get the following output:

```
error.cpp:6:15: error: out-of-line definition of 'doSometing' ...
[clang-diagnostic-error]
void MyClass::doSometing() {}
              ^~~~~~~~~~
              doSomething
error.cpp:3:8: note: 'doSomething' declared here
  void doSomething();
       ^
Found compiler error(s).
```

Figure 5.34: Running a fixed misc-classchecker check on a file containing compilation errors

As we can see, the compiler error is reported, but our check does not trigger any warnings.

Despite the fact that we correctly processed the unusual clang-tidy result, it's worth noting that not every compilation error can be correctly processed. As mentioned in *Section 3.7, Processing AST in the case of errors*, the Clang compiler tries to produce an AST even when encountering compilation errors. This approach is because it's designed for use by IDEs and other tools that benefit from as much information as possible, even in the presence of errors. However, this "error-recovery" mode of the AST can produce structures that Clang-Tidy might not anticipate. Therefore, we should adhere to the following rule:

> **Tip**
>
> Always ensure your code compiles without errors before running Clang-Tidy and other Clang Tools. This guarantees that the AST is both accurate and complete.

5.5   Summary

In this chapter, we delved into Clang-Tidy, a robust tool for code analysis. We explored its configuration, execution, and internal architecture. Additionally, we developed a custom Clang-Tidy check to assess class complexity. Our check utilized basic AST matchers, akin to regular expressions within the AST. For complexity determination, we employed a simple method. More sophisticated metrics, such as cyclomatic complexity, demand tools such as **Control Flow Graphs (CFGs)**. The adventure continues in the next chapter, where we'll dive deep into designing intricate checks using CFG.

5.6   Further reading

- Clang-Tidy extra Clang tools documentation: `https://clang.llvm.org/extra/clang-tidy/`

- AST matcher reference: `https://clang.llvm.org/docs/LibASTMatchersReference.html`

6

Advanced Code Analysis

Clang-Tidy checks, as discussed in the previous chapter, rely on advanced matching provided by the AST. However, this approach might not be sufficient for detecting more complex problems, such as lifetime issues (that is, when an object or resource is accessed or referenced after it has been deallocated or has gone out of scope, potentially leading to unpredictable behavior or crashes). In this chapter, we will introduce advanced code analysis tools based on the **Control Flow Graph (CFG)**. The Clang Static Analyzer is an excellent example of such tools, and Clang-Tidy also integrates some aspects of CFGs. We will begin with typical usage examples and then delve into the implementation details. The chapter will conclude with a custom check that employs advanced techniques and extends the concept of class complexity to method implementations. We will define cyclomatic complexity and demonstrate how to calculate it using the CFG library provided by Clang. In this chapter, we will explore the following topics:

- What static analysis is

- Gaining knowledge of CFGs – the basic data structure used for static analysis

- How CFGs can be used in a custom Clang-Tidy check

- What analysis tools are provided in Clang and what are their limitations

6.1   Technical requirements

The source code for this chapter is located in the `chapter6` folder of the book's GitHub repository: `https://github.com/PacktPublishing/Clang-Compiler-Frontend-Packt /tree/main/chapter6`.

6.2   Static analysis

Static analysis is a crucial technique in software development that involves inspecting the code without actually running the program. This method focuses on analyzing either the source code or its compiled version to detect a variety of issues, such as errors, vulnerabilities, and deviations from coding standards. Unlike dynamic analysis, which requires the execution of the program, static analysis allows for examining the code in a non-runtime environment.

More generally, static analysis aims to check a specific property of a computer program based on its meaning; that is, it can be considered a part of semantic analysis (see *Section 2.2.2, Parser*). For instance, if C is the set of all C/C++ programs and P is a property of such a program, then the goal of static analysis is to check the property for a specific program $P \in C$, that is, to answer the question of whether $P(P)$ is true or false.

Our Clang-Tidy check from the previous chapter (see *Section 5.4, Custom Clang-Tidy check*) is a good example of such a property. In reality, it takes C++ code with a class definition and decides whether the class is complex or not based on the number of methods it has.

It's worth noting that not all properties of a program can be checked. The most obvious example is the famous halting problem [31].

> **Important note**
>
> The halting problem can be formulated as follows: Given a program P and an input I, determine whether P halts or continues to run indefinitely when executed with I.
>
> Formally, the problem is to decide, for a given program P and an input I, whether the computation of $P(I)$ eventually stops (halts) or will never terminate (loops indefinitely).
>
> Alan Turing proved that there is no general algorithmic method for solving this problem for all possible program-input pairs. This result implies that there is no single algorithm that can correctly determine for every pair (P, I) whether P halts when run with I.

Despite the fact that not all properties of programs can be proven, it can be done for some cases. There is a reasonable number of such cases that make static analysis a practical tool for usage. Thus, we can use the tools in these cases to systematically scan the code to determine properties of the code. These tools are adept at identifying issues ranging from simple syntax errors to more complex potential bugs. One of the key benefits of static analysis is its ability to catch problems early in the development cycle. This early detection is not only efficient but also resource-saving, as it helps identify and rectify issues before the software is run or deployed.

Static analysis plays a significant role in ensuring the quality and compliance of software. It checks that the code adheres to prescribed coding standards and guidelines, which is particularly important in large-scale projects or industries with strict regulatory requirements. Moreover, it is highly effective in uncovering common security vulnerabilities such as buffer overflows, SQL injection flaws, and cross-site scripting vulnerabilities.

Additionally, static analysis contributes to code refactoring and optimization by pinpointing areas of redundancy, unnecessary complexity, and opportunities for improvement. It's a common practice to integrate these tools into the development process, including continuous integration pipelines. This integration allows for ongoing analysis of the code with each new commit or build, ensuring continual quality assurance.

The Clang-Tidy checks that we created in the last chapter can be considered an example of a static analysis program. In this chapter, we will consider more advanced topics involving data structures such as CFGs, which we will see next.

6.3 CFG

A **CFG** is a fundamental data structure in compiler design and static program analysis, representing all paths that might be traversed through a program during execution.

A CFG consists of the following key components:

- **Nodes**: Correspond to basic blocks, a straight-line sequence of operations with one entry and one exit point

- **Edges**: Represent the flow of control from one block to another, including both conditional and unconditional branches

- **Start and end nodes**: Every CFG has a unique entry node and one or more exit nodes

As an example of a CFG, consider the function to calculate the maximum of two integer numbers that we used as an example before; see Figure 2.5:

```
1 int max(int a, int b) {
2   if (a > b)
3     return a;
4   return b;
5 }
```

Figure 6.1: CFG example C++ code: max.cpp

The corresponding CFG can be represented as follows:

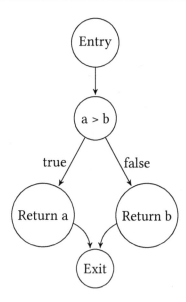

Figure 6.2: CFG example for max.cpp

The diagram shown in Figure 6.2 visually represents a CFG for the max function (from Figure 6.1) with a series of connected nodes and directed edges:

- **Entry node**: At the top, there is an "**entry**" node, representing the starting point of the function's execution.

- **Conditional node**: Below the entry node, there is a node labeled "**a > b**". This node represents the conditional statement in the function, where the comparison between *a* and *b* is made.

- **Branches for true and false conditions**:

    - On the true branch (left side), there is a node labeled "**Return a**", connected by an edge from the "**a > b**" node. This edge is labeled "**true**", indicating that if *a* is greater than *b*, the flow goes to this node.

    - On the false branch (right side), there is a node labeled "**Return b**", connected by an edge from the "**a > b**" node. This edge is labeled "**false**", indicating that if *a* is not greater than *b*, the flow goes to this node.

- **Exit node**: Below both the "**Return a**" and "**Return b**" nodes, converging at a point, there is an "**exit**" node. This represents the termination point of the function, where the control flow exits the function after returning either a or b.

This CFG effectively illustrates how the max function processes input and reaches a decision on which value to return based on the comparison.

The CFG representation can also be used to estimate function complexity. In brief, a more complex picture corresponds to a more complex system. We will use a precise definition of complexity known as cyclomatic complexity, or M [28], which can be calculated as follows:

$$M = E - N + 2P \tag{6.1}$$

where:

- E is the number of edges in the graph

- N is the number of nodes in the graph

- P is the number of connected components (for a single CFG, P is usually 1)

For the max function discussed earlier, the CFG can be analyzed as follows:

- **Nodes (N)**: There are five nodes (Entry, $a > b$, Return a, Return b, Exit)

- **Edges (E)**: There are five edges (from Entry to $a > b$, from $a > b$ to Return a, from $a > b$ to Return b, from Return a to Exit, and from Return b to Exit)

- **Connected components (P)**: As it's a single function, $P = 1$

Substituting these values into the formula, we get the following:

$$M = 5 - 5 + 2 \times 1 = 2$$

Thus, the cyclomatic complexity of the max function, based on the given CFG, is 2. This indicates that there are two linearly independent paths through the code, corresponding to the two branches of the if statement.

Our next step will be to create a Clang-Tidy check that uses a CFG to calculate cyclomatic complexity.

6.4 Custom CFG check

We are going to use the knowledge gained in *Section 5.4, Custom Clang-Tidy check* to create a custom CFG check. As mentioned previously, the check will use Clang's CFG to calculate cyclomatic complexity. The check should issue a warning if the calculated complexity exceeds a threshold. This threshold will be set up as a configuration parameter, allowing us to change it during our tests. Let's start with the creation of the project skeleton.

6.4.1 Creating the project skeleton

We will use `cyclomaticcomplexity` as the name for our check, and our project skeleton can be created as follows:

```
$ ./clang-tools-extra/clang-tidy/add_new_check.py misc cyclomaticcomplexity
```

Figure 6.3: Creating a skeleton for the misc-cyclomaticcomplexity check

As a result of the run, we will get a number of modified and new files. The most important ones for us are the following two files located in the `clang-tools-extra/clang-tidy/misc/` folder:

- `misc/CyclomaticcomplexityCheck.h` : This is the header file for our check

- `misc/CyclomaticcomplexityCheck.cpp` : This file will house the implementation of our check

These files need to be modified to achieve the required functionality for the check.

6.4.2   Check implementation

For the header file, we aim to add a private function to calculate the cyclomatic complexity. Specifically, the following code needs to be inserted:

```
27 private:
28   unsigned calculateCyclomaticComplexity(const CFG *cfg);
```

Figure 6.4: Modifications to CyclomaticcomplexityCheck.h

More substantial modifications are required in the `.cpp` file. We will begin with the implementation of the `registerMatchers` method, as follows:

```
17 void CyclomaticcomplexityCheck::registerMatchers(MatchFinder *Finder) {
18   Finder->addMatcher(functionDecl().bind("func"), this);
19 }
```

Figure 6.5: Modifications to CyclomaticcomplexityCheck.cpp: registerMatchers implementation

Based on the code, our check will be applied only to function declarations, `clang::FunctionDecl`. The code can also be extended to support other C++ constructs.

The implementation of the `check` method is presented in Figure 6.6. At *Lines 22-23*, we perform basic checks on the matched AST node, `clang::FunctionDecl` in our case. At *Lines 25-26*, we create the CFG object using the `CFG::buildCFG` method. The first two parameters specify the declaration (`clang::Decl`) and the statement for the declaration (`clang::Stmt`). At *Line 30*, we calculate the cyclomatic complexity using the threshold, which can be obtained as the `"Threshold"` option of our check. This provides flexibility in testing for different input programs. *Lines 31-34* contain the implementation of the check result printout.

```
21 void CyclomaticcomplexityCheck::check(const MatchFinder::MatchResult
   ↪ &Result) {
22   const auto *Func = Result.Nodes.getNodeAs<FunctionDecl>("func");
23   if (!Func || !Func->hasBody()) return;
24
25   std::unique_ptr<CFG> cfg =
26       CFG::buildCFG(Func, Func->getBody(), Result.Context,
          ↪ CFG::BuildOptions());
27   if (!cfg) return;
28
29   unsigned Threshold = Options.get("Threshold", 5);
30   unsigned complexity = calculateCyclomaticComplexity(cfg.get());
31   if (complexity > Threshold) {
32     diag(Func->getLocation(), "function %0 has high cyclomatic complexity
        ↪ (%1)")
33         << Func << complexity;
34   }
35 }
```

Figure 6.6: Modifications to CyclomaticcomplexityCheck.cpp: check implementation

The calculateCyclomaticComplexity method is used to calculate the cyclomatic complexity. It takes the created clang::CFG object as an input parameter. The implementation is shown in the following figure:

```
37 unsigned CyclomaticcomplexityCheck::calculateCyclomaticComplexity(
38     const CFG *cfg) {
39   unsigned edges = 0;
40   unsigned nodes = 0;
41
42   for (const auto *block : *cfg) {
43     edges += block->succ_size();
44     ++nodes;
45   }
46
47   return edges - nodes + 2;  // Simplified formula
48 }
```

Figure 6.7: Modifications to CyclomaticcomplexityCheck.cpp: calculateCyclomaticComplexity implementation

We iterate over all CFG blocks at *Lines 42-45*. The number of blocks corresponds to the number of nodes, denoted as N in Equation (6.1). We sum up the number of successors for each block to calculate the number of edges, denoted as E. We assume that the number of connected components, denoted as P, is equal to one for our simplified example.

After implementing the check, it's time to build and run our new check on our example; see Figure 6.1.

6.4.3   Building and testing the cyclomatic complexity check

We will use the basic build configuration specified in Figure 1.4 and build Clang-Tidy using the standard command from Figure 5.2:

```
$ ninja install-clang-tidy
```

Assuming the build configuration from Figure 1.4, this command will install the Clang-Tidy binary in the `<...>/llvm-project/install/bin` folder.

> **Important note**
>
> If you use a build configuration with shared libraries (with the
> BUILD_SHARED_LIBS flag set to ON), as shown in Figure 1.12, then you might need
> to install and built all artifacts with `ninja install` .

We will run our check on the example program shown in Figure 6.1. As we previously
calculated, the cyclomatic complexity for the test is 2, which is lower than the default value
of 5 specified at *Line 29* in our check method implementation, as seen in Figure 6.6. Thus,
we need to override the default value to 1 to be able to see a warning in our test program.
This can be done using the -config option, which we previously used for classchecker
check tests, as shown in Figure 5.20. The command for the test will look as follows:

```
1 $ <...>/llvm-project/install/bin/clang-tidy                              \
2    -checks="-*,misc-cyclomaticcomplexity"                                \
3    -config="{CheckOptions:                                               \
4              [{key: misc-cyclomaticcomplexity.Threshold, value: '1'}]}" \
5    max.cpp                                                               \
6    -- -std=c++17
```

Figure 6.8: Testing cyclomatic complexity on the max.cpp example

Line 2 in Figure 6.8 indicates that we want to run only one Clang-Tidy check:
misc-cyclomaticcomplexity . At *lines 3-4*, we set up the required threshold. *Line 5*
specifies the name of the file being tested (max.cpp in our case), and the final line, *Line 6*,
contains some compilation flags for our program.

We will get the following output if we run the command from Figure 6.8:

```
max.cpp:1:5: warning: function 'max' has high cyclomatic complexity (2) ...
int max(int a, int b) {
    ^
```

Figure 6.9: Testing cyclomatic complexity on the max.cpp example: output

The following question might arise: How does Clang build the CFG? We can use a debugger to investigate the process.

6.5   CFG on Clang

A CFG is the basic data structure for advanced static analysis using Clang tools. Clang constructs the CFG for a function from its AST, identifying basic blocks and control flow edges. Clang's CFG construction handles various C/C++ constructs, including loops, conditional statements, switch cases, and complex constructs such as `setjmp/longjmp` and C++ exceptions. Let's consider the process using our example from Figure 6.1.

6.5.1   CFG construction by example

Our example from Figure 6.1 has five nodes, as shown in Figure 6.2. Lets run a debugger to investigate the process, as follows:

```
1  $ lldb <...>/llvm-project/install/bin/clang-tidy --              \
2    -checks="-*,misc-cyclomaticcomplexity"                         \
3    -config="{CheckOptions:                                        \
4              [{key: misc-cyclomaticcomplexity.Threshold, value: '1'}]}" \
5    max.cpp                                                        \
6    -- -std=c++17 -Wno-all
```

Figure 6.10: Debugger session running to investigate the CFG creation process

We used the same command as in Figure 6.8 but changed the first line of the command to run the check via a debugger. We also changed the last line to suppress all warnings from the compiler.

Important note

Advanced static analysis is a part of semantic analysis. For example, warnings are printed if Clang detects unreachable code, controlled by the -Wunreachable-code option. The detector is a part of Clang's semantic analysis and utilizes CFGs, in addition to ASTs, as the basic data structures to detect such issues. We can suppress these warnings and, as a result, disable CFG initialization in Clang by specifying the special -Wno-all command-line option, which suppresses all warnings generated by the compiler.

We will set a breakpoint at the CFGBuilder::createBlock function, which creates a CFG block.

```
$ lldb <...>/llvm-project/install/bin/clang-tidy --              \
  -checks="-*,misc-cyclomaticcomplexity"                         \
  -config="{CheckOptions:                                        \
            [{key: misc-cyclomaticcomplexity.Threshold, value: '1'}]}" \
  max.cpp                                                        \
  -- -std=c++17 -Wno-all
...
(lldb) b CFGBuilder::createBlock
Breakpoint 1: where = ...CFGBuilder::createBlock(bool) const ...
```

Figure 6.11: Running debugger and setting breakpoint for CFGBuilder::createBlock

If we run the debugger, we will see that the function is called five times for our example; that is, five CFG blocks are created for our max function:

```
 1 (lldb) r
 2 ...
 3     frame #0: ...CFGBuilder::createBlock...
 4   1690 /// createBlock - Used to lazily create blocks that are connected
 5   1691 ///  to the current (global) successor.
 6   1692 CFGBlock *CFGBuilder::createBlock(bool add_successor) {
 7 -> 1693   CFGBlock *B = cfg->createBlock();
 8   1694   if (add_successor && Succ)
 9   1695     addSuccessor(B, Succ);
10   1696   return B;
11
12 (lldb) c
13 ...
14 (lldb) c
15 ...
16 (lldb) c
17 ...
18 (lldb) c
19 ...
20 (lldb) c
21 ...
22 1 warning generated.
23 max.cpp:1:5: warning: function 'max' has high cyclomatic complexity (2)
    ↪ [misc-cyclomaticcomplexity]
24 int max(int a, int b) {
25     ^
26 Process ... exited with status = 0 (0x00000000)
```

Figure 6.12: Creation of CFG blocks, with breakpoints highlighted

The debugger session shown in Figure 6.12 can be considered the entry point to the CFG creation process. Now, it's time to delve deeply into the implementation details.

6.5.2 CFG construction implementation details

The blocks are created in reverse order, as seen in Figure 6.13. The first block to be created is the exit block, as shown in Figure 6.13, *Line 4*. Then, the CFG builder traverses the `clang::Stmt` object passed as a parameter (*Line 9*). The entry block is created last, at *Line 12*:

```
1  std::unique_ptr<CFG> CFGBuilder::buildCFG(const Decl *D, Stmt *Statement) {
2    ...
3    // Create an empty block that will serve as the exit block for the CFG.
4    Succ = createBlock();
5    assert(Succ == &cfg->getExit());
6    Block = nullptr;  // the EXIT block is empty.  ...
7    ...
8    // Visit the statements and create the CFG.
9    CFGBlock *B = Visit(Statement, ...);
10   ...
11   // Create an empty entry block that has no predecessors.
12   cfg->setEntry(createBlock());
13   ...
14   return std::move(cfg);
15 }
```

Figure 6.13: Simplified buildCFG implementation from clang/lib/Analysis/CFG.cpp

The visitor uses the `clang::Stmt::getStmtClass` method to implement an ad hoc visitor based on the type of the statement, as shown in the following code snippet:

```
 1 CFGBlock *CFGBuilder::Visit(Stmt * S, ...) {
 2   ...
 3   switch (S->getStmtClass()) {
 4     ...
 5     case Stmt::CompoundStmtClass:
 6       return VisitCompoundStmt(cast<CompoundStmt>(S), ...);
 7     ...
 8     case Stmt::IfStmtClass:
 9       return VisitIfStmt(cast<IfStmt>(S));
10     ...
11     case Stmt::ReturnStmtClass:
12     ...
13       return VisitReturnStmt(S);
14     ...
15   }
16 }
```

Figure 6.14: Statement visitor implementation; the cases used for our example are highlighted, the code was taken from clang/lib/Analysis/CFG.cpp

Our example includes two return statements and one if statement, which are combined into a compound statement. The relevant parts of the visitor are shown in Figure 6.14.

In our case, the passed statement is a compound statement; therefore, *Line 6* from Figure 6.14 is activated. The following code is then executed:

```
1 CFGBlock *CFGBuilder::VisitCompoundStmt(CompoundStmt *C, ...) {
2   ...
3   CFGBlock *LastBlock = Block;
4
5   for (Stmt *S : llvm::reverse(C->body())) {
6     // If we hit a segment of code just containing ';' (NullStmts), we can
7     // get a null block back.  In such cases, just use the LastBlock
8     CFGBlock *newBlock = Visit(S, ...);
9
10     if (newBlock)
11       LastBlock = newBlock;
12
13     if (badCFG)
14       return nullptr;
15     ...
16   }
17
18   return LastBlock;
19 }
```

Figure 6.15: Compound statement visitor, the code was taken from clang/lib/Analysis/CFG.cpp

Several constructions are visited while the CFG is being created for our example. The first one is clang::IfStmt. The relevant parts are shown in the following figure:

```
1  CFGBlock *CFGBuilder::VisitIfStmt(IfStmt *I) {
2    ...
3    // Process the true branch.
4    CFGBlock *ThenBlock;
5    {
6      Stmt *Then = I->getThen();
7      ...
8      ThenBlock = Visit(Then, ...);
9      ...
10   }
11
12   // Specially handle "if (expr1 || ...)" and "if (expr1 && ...)"
13   // ...
14   if (Cond && Cond->isLogicalOp())
15     ...
16   else {
17     // Now create a new block containing the if statement.
18     Block = createBlock(false);
19     ...
20   }
21   ...
22 }
```

Figure 6.16: If statement visitor, the code was taken from clang/lib/Analysis/CFG.cpp

A special block for the `if` statement is created at *Line 18*. We also visit the 'then' condition at *Line 8*.

The 'then' condition leads to visiting a return statement. The corresponding code is as follows:

```
1  CFGBlock *CFGBuilder::VisitReturnStmt(Stmt *S) {
2    // Create the new block.
3    Block = createBlock(false);
4    ...
5    // Visit children
6    if (ReturnStmt *RS = dyn_cast<ReturnStmt>(S)) {
7      if (Expr *O = RS->getRetValue())
8        return Visit(O, ...);
9      return Block;
10   }
11   ...
12 }
```

Figure 6.17: Return statement visitor, the code was taken from clang/lib/Analysis/CFG.cpp

For our example, it creates a block at *Line 3* and visits the return expression at *Line 8*. Our return expression is a trivial one that does not necessitate the creation of a new block.

The code fragments presented in Figure 6.13 to Figure 6.17 show only the block creation procedure. Some important parts were omitted for simplicity. Notably, the build procedure also involves the following:

- Edge creation: A typical block can have one or more successors. The list of nodes (blocks) with a list of successors (edges) for each block maintains the entire graph structure, representing symbolic program execution.

- Storing meta-information: Each block stores additional meta-information associated with it. For instance, each block keeps a list of statements in the block.

- Processing edge cases: C++ is a complex language with many different language constructs that require special processing.

The CFG is a fundamental data structure for advanced code analysis. Clang has several tools created using CFGs. Let's briefly look at them.

6.6   Brief description of Clang analysis tools

As mentioned earlier, the CFG is foundational for other analysis tools in Clang, several of which have been created atop the CFG. These tools also employ advanced mathematics to analyze various cases. The most notable tools are as follows [32]:

- LivenessAnalysis: Determines whether a computed value will be used before being overwritten, producing liveness sets for each statement and CFGBlock

- UninitializedVariables: Identifies the use of uninitialized variables through multiple passes, including initial categorization of statements and subsequent calculation of variable usages

- Thread Safety Analysis: Analyzes annotated functions and variables to ensure thread safety

LivenessAnalysis in Clang is essential for optimizing code by determining whether a value computed at one point will be used before being overwritten. It produces liveness sets for each statement and CFGBlock, indicating potential future use of variables or expressions. This backward "may" analysis simplifies read/write categorization by treating variable declarations and assignments as writes, and other contexts as reads, regardless of aliasing or field usage. Valuable in dead code elimination and compiler optimizations, such as efficient register allocation, it helps free up memory resources and improve program efficiency. Despite challenges with corner cases and documentation, its straightforward implementation and the ability to cache and query results make it a vital tool in enhancing software performance and resource management.

> **Important note**
>
> Forward analysis is a method used in programming to check how data moves through a program from start to finish. Following the data path step by step as the program runs allows us to see how it changes or where it goes. This method is instrumental for identifying issues such as improperly set-up variables or tracking data flow in the program. It contrasts with backward analysis, which starts at the end of the program and works backward.

UninitializedVariables analysis in Clang, designed to detect the use of variables before initialization, operates as a forward "must" analysis. It involves multiple passes, including initial code scanning for statement classification and subsequent use of a fix-point algorithm to propagate information through the CFG. Handling more sophisticated scenarios than LivenessAnalysis, it faces challenges such as lacking support for record fields and non-reusable analysis results, limiting its efficiency in certain situations.

Thread Safety Analysis in Clang, a forward analysis, focuses on ensuring proper synchronization in multithreaded code. It computes sets of locked mutexes for each statement in a block and utilizes annotations to indicate guarded variables or functions. Translating Clang expressions into TIL (Typed Intermediate Language)[32], it effectively handles the complexity of C++ expressions and annotations. Despite strong C++ support and a sophisticated understanding of variable interactions, it faces limitations, such as lack of support for aliasing, which can lead to false positives.

6.7 Knowing the limitations of analysis

It's worth mentioning some limitations of the analysis that can be conducted with Clang's AST and CFG. The most notable ones are mentioned here [2]:

- Limitations of Clang's AST: Clang's AST is unsuitable for data flow analysis and control flow reasoning, leading to inaccurate results and inefficient analysis due to the loss of vital language information. Soundness of analysis is also a consideration, where the precision of certain analyses, such as liveness analysis, can be valuable if they are precise enough rather than always being conservative.

- Issues with Clang's CFG: While Clang's CFG aims to bridge the gap between AST and LLVM IR, it encounters known problems, has limited interprocedural capabilities, and lacks adequate testing coverage.

One example mentioned in [2] relates to C++ coroutines, a new feature introduced in C++20. Some aspects of this functionality are implemented outside the Clang frontend and are not visible with tools such as Clang's AST and CFG. This limitation makes analysis, especially lifetime analysis, tricky for such functionalities.

Despite these limitations, Clang's CFG remains a powerful tool widely used in compiler and compiler tool development. There is also active development of other tools [27] that aim to close the gaps in Clang's CFG capabilities.

6.8 Summary

In this chapter, we investigated Clang's CFG, a powerful data structure that represents the symbolic execution of a program. We created a simple Clang-Tidy check using a CFG to calculate cyclomatic complexity, a metric useful for estimating code complexity. Additionally, we explored the details of CFG creation and the formation of its basic internal structures. We discussed some tools developed with CFGs, which are useful for detecting lifetime issues, thread safety, and uninitialized variables. We also briefly described the limitations of CFGs and how other tools can address these limitations.

The next chapter will cover refactoring tools. These tools can perform complex code modifications using the AST provided by the Clang compiler.

6.9 Future reading

- Flemming Nielson, Hanne Riis Nielson, and Chris Hankin, *Principles of Program Analysis*, Springer, 2005 [29]

- Xavier Rival and Kwangkeun Yi, *Introduction to Static Analysis: An Abstract Interpretation Perspective*, The MIT Press, 2020 [30]

- Kristóf Umann *A survey of dataflow analyses in Clang*: `https://lists.llvm.org/pipermail/cfe-dev/2020-October/066937.html`

- Bruno Cardoso Lopes and Nathan Lanza *An MLIR based Clang IR (CIR)*: `https://discourse.llvm.org/t/rfc-an-mlir-based-clang-ir-cir/63319`

7

Refactoring Tools

Clang is renowned for its ability to provide suggestions for code fixes. For instance, if you miss a semicolon, Clang will suggest that you insert it. The ability to modify source code goes beyond the compilation process and is widely used in various tools for code modifications, particularly in refactoring tools. The ability to offer fixes is a powerful feature that extends the capabilities of a linter framework, such as Clang-Tidy, which not only detects issues but also provides suggestions for fixing them.

In this chapter, we will explore refactoring tools. We will begin by discussing the fundamental classes used for code modification, notably `clang::Rewriter`. We will use Rewriter to build a custom refactoring tool that changes method names within a class. Later in the chapter, we will reimplement the tool using Clang-Tidy and delve into `clang::FixItHint`, a component of the Clang Diagnostics subsystem that is employed by both Clang-Tidy and the Clang compiler to modify source code.

To conclude the chapter, we will introduce a crucial Clang tool called Clang-Format. This tool is widely employed for code formatting. We will explore the functionality offered by the tool, delve into its design, and understand the rationale behind specific design decisions made during its development.

The chapter covers the following topics:

- How to create a custom Clang tool for code refactoring

- How to integrate code modifications into a Clang-Tidy check

- An overview of Clang-Format and how it can be integrated with Clang-Tidy

7.1   Technical requirements

The source code for this chapter is located in the `chapter7` folder of the book's GitHub repository: `https://github.com/PacktPublishing/Clang-Compiler-Frontend-Packt /tree/main/chapter7`.

7.2   Custom code modification tool

We will create a Clang tool that will help us to rename methods for a class that is used for unit testing. We will start with a description for the `clang::Rewriter` class – the basic class that is used for code modifications.

7.2.1   Code modification support at Clang

`clang::Rewriter` is a Clang library class that facilitates source code rewriting operations within a translation unit. It provides methods for inserting, removing, and replacing code within the **Abstract Syntax Tree (AST)** of the source code. Developers can use `clang::Rewriter` for complex code modifications, such as restructuring or generating new code constructs. It can be applied for both code generation and code refactoring tasks, making it versatile for various code transformation purposes.

The class has several methods for text insertion; for instance, `clang::Rewriter ::InsertText` inserts the text at the specified source location, and `clang ::SourceLocation` is used to specify the exact location at the buffer, see *Section 4.4.1, SourceManager and SourceLocation.* In addition to the text insertion, you can also remove text with clang::Rewriter::RemoveText or replace text with a new one using clang::Rewriter::ReplaceText. The last two use source range (clang::SourceRange) to specify the positions at the text to be removed or replaced.

clang::Rewriter uses clang::SourceManager, as explained in *Section 4.4.1, SourceManager and SourceLocation*, to access the source code that needs to be modified. Let's look at how Rewriter can be used in a real project.

7.2.2 Test class

Suppose we have a class that is used for tests. The class name starts with the "Test" prefix (for instance, TestClass), but there aren't any 'test_' prefixes for public methods of the class. For instance, the class has a public method with the name 'pos' (TestClass::pos) instead of 'test_pos' (TestClass::test_pos()). We want to create a tool that will add such a prefix for the class methods.

```
1 class TestClass {
2 public:
3    TestClass(){};
4    void pos(){};
5
6 private:
7    void private_pos(){};
8 };
```

```
1 class TestClass {
2 public:
3    TestClass(){};
4    void test_pos(){};
5
6 private:
7    void private_pos(){};
8 };
```

Original code   *Modified code*

Figure 7.1: Code transformations for TestClass

Thus, we want the method TestClass::pos (see Figure 7.1) to be replaced with TestClass::test_pos at the class declaration.

If we have a code where we make a call to the method, the following replacement should be made:

1	`TestClass test;`
2	`test.pos();`

1	`TestClass test;`
2	`test.test_pos();`

Original code   *Modified code*

Figure 7.2: Code transformations for TestClass's method calls

The tool should also ignore all public methods with the required modifications already applied, either manually or automatically. In other words, if a method already has the required 'test_' prefix, the tool should not modify it.

We are going to create a Clang tool called 'methodrename', which will perform all the required code modifications. This tool will utilize the recursive AST visitor discussed in *Section 3.4, Recursive AST visitor*. The most crucial aspect is the implementation of the Visitor class. Let's examine it in detail.

7.2.3   Visitor class implementation

Our Visitor class should handle specific processing for the following AST nodes:

- `clang::CXXRecordDecl`: This involves processing C++ class definitions with names starting with the "Test" prefix. For such classes, all user-defined public methods should be prefixed with "test_".

- `clang::CXXMemberCallExpr`: Additionally, we need to identify all instances where the modified method is used and make the corresponding changes following the method's renaming in the class definition.

The processing for clang::CXXRecordDecl nodes will be as follows:

```
10    bool VisitCXXRecordDecl(clang::CXXRecordDecl *Class) {
11      if (!Class->isClass())
12        return true;
13      if (!Class->isThisDeclarationADefinition())
14        return true;
15      if (!Class->getName().starts_with("Test"))
16        return true;
17      for (const clang::CXXMethodDecl *Method : Class->methods()) {
18        clang::SourceLocation StartLoc = Method->getLocation();
19        if (!processMethod(Method, StartLoc, "Renamed method"))
20          return false;
21      }
22      return true;
23    }
```

Figure 7.3: CXXRecordDecl visitor implementation

Lines 11-16 in Figure 7.3 represent the conditions that we require from the examined node. For example, the corresponding class name should start with the "Test" prefix (see *Lines 15-16* in Figure 7.3), where we utilize the starts_with() method of the llvm::StringRef class.

After verifying these conditions, we proceed to examine the methods within the found class.

The verification process is implemented in the Visitor::processMethod method, and its implementation is presented in the following code fragment:

```
44    bool processMethod(const clang::CXXMethodDecl *Method,
45                        clang::SourceLocation StartLoc, const char
        ↪  *LogMessage) {
46      if (Method->getAccess() != clang::AS_public)
47        return true;
48      if (llvm::isa<clang::CXXConstructorDecl>(Method))
49        return true;
50      if (!Method->getIdentifier() || Method->getName().starts_with("test_"))
51        return true;
52
53      std::string OldMethodName = Method->getNameAsString();
54      std::string NewMethodName = "test_" + OldMethodName;
55      clang::SourceManager &SM = Context.getSourceManager();
56      clang::tooling::Replacement Replace(SM, StartLoc,
        ↪  OldMethodName.length(),
57                                          NewMethodName);
58      Replaces.push_back(Replace);
59      llvm::outs() << LogMessage << ": " << OldMethodName << " to "
60                   << NewMethodName << "\n";
61      return true;
62    }
```

Figure 7.4: Implementation of processMethod

Lines 46-51 in Figure 7.4 contain the checks for the required conditions. For instance, in *Lines 46-47*, we verify that the method is public. *Lines 48-49* are used to exclude constructors from processing, and *Lines 50-51* serve to exclude methods that already have the required prefix.

The main replacement logic is implemented in *Lines 53-58*. Particularly, in *Lines 56-57*, we create a special clang::tooling::Replacement object, which serves as a wrapper for required code modifications. The object's parameters are as follows:

1. clang::SourceManager: We obtain the source manager from clang::ASTContext at *Line 55*.

2. clang::SourceLocation: The source location specifies the starting position for replacement. The position is passed as the second parameter of our processMethod method, as seen in *Line 45*.

3. unsigned: The length of the replaced text.

4. clang::StringRef: The replacement text, which we create at *Line 54*.

We store the replacement in the Replaces object, a private member of our Visitor class:

```
40 private:
41    clang::ASTContext &Context;
42    std::vector<clang::tooling::Replacement> Replaces;
```

There is a special getter to access the object outside the Visitor class:

```
36    const std::vector<clang::tooling::Replacement> &getReplacements() {
37       return Replaces;
38    }
```

We log the action at *Lines 59-60*, using `LogMessage` as the prefix for the log message. Different log messages are used for different AST nodes; for instance, we use "Renamed method" (see Figure 7.3, *Line 19*) for `clang::CXXRecordDecl`.

The log message will be different for the method call. The corresponding processing is shown in the following figure.

```
25   bool VisitCXXMemberCallExpr(clang::CXXMemberCallExpr *Call) {
26     if (clang::CXXMethodDecl *Method = Call->getMethodDecl()) {
27       clang::CXXRecordDecl *Class = Method->getParent();
28       if (!Class->getName().starts_with("Test"))
29         return true;
30       clang::SourceLocation StartLoc = Call->getExprLoc();
31       return processMethod(Method, StartLoc, "Renamed method call");
32     }
33     return true;
34   }
```

Figure 7.5: CXXMemberCallExpr visitor implementation

We verify that the class name, which holds the test method, starts with the 'Test' prefix at *Lines 27-29*. The replacement source location is obtained at *Line 30*. At *Line 31*, we call our `processMethod` function to process the found method, passing the "Renamed method call" as the log message to the call.

The `Visitor` is initialized in the `Consumer` class, which will be our next goal.

7.2.4   Consumer class implementation

The Consumer class initializes the Visitor and starts AST traversal in the HandleTranslationUnit method. The class can be written as follows:

```
6  class Consumer : public clang::ASTConsumer {
7  public:
8    void HandleTranslationUnit(clang::ASTContext &Context) override {
9      Visitor V(Context);
10     V.TraverseDecl(Context.getTranslationUnitDecl());
11
12     // Apply the replacements.
13     clang::Rewriter Rewrite(Context.getSourceManager(),
        ↪ clang::LangOptions());
14     auto &Replaces = V.getReplacements();
15     for (const auto &Replace : Replaces) {
16       if (Replace.isApplicable()) {
17         Replace.apply(Rewrite);
18       }
19     }
20
21     // Apply the Rewriter changes.
22     if (Rewrite.overwriteChangedFiles()) {
23       llvm::errs() << "Error: Cannot apply changes to the file\n";
24     }
25   }
26 };
27 } // namespace methodrename
```

Figure 7.6: Consumer class implementation

We initialize the Visitor and begin traversal at *Lines 9-10* (see Figure 7.6). The Rewriter is created at *Line 13*, and replacements are applied at *Lines 14-19*. Finally, the result is stored in the original file at *Lines 22-24*.

The Visitor and Consumer classes are wrapped within the clangbook::methodrename namespace. The Consumer instance is created in the FrontendAction class. This class's implementation mirrors that of the RecursiveVisitor and DeclVisitor, as detailed in Figure 3.8. The only difference is the use of the clangbook::methodrename namespace for the new tool.

7.2.5   Build configuration and main function

The main function for our tool is similar to the recursive visitor one defined in Figure 3.21:

```
13 int main(int argc, const char **argv) {
14   llvm::Expected<clang::tooling::CommonOptionsParser> OptionsParser =
15       clang::tooling::CommonOptionsParser::create(argc, argv,
         ↪ TestCategory);
16   if (!OptionsParser) {
17     llvm::errs() << OptionsParser.takeError();
18     return 1;
19   }
20   clang::tooling::ClangTool Tool(OptionsParser->getCompilations(),
21                                 OptionsParser->getSourcePathList());
22   return Tool.run(clang::tooling::newFrontendActionFactory<
23                   clangbook::methodrename::FrontendAction>()
24                   .get());
25 }
```

Figure 7.7: The main function for the 'methodrename' test tool

As you can see, we changed only the namespace name for our custom frontend action at *Line 23.*

The build configuration is specified as follows:

```
1  cmake_minimum_required(VERSION 3.16)
2  project("methodrename")
3
4  if ( NOT DEFINED ENV{LLVM_HOME})
5    message(FATAL_ERROR "$LLVM_HOME is not defined")
6  else()
7    message(STATUS "$LLVM_HOME found: $ENV{LLVM_HOME}")
8    set(LLVM_HOME $ENV{LLVM_HOME} CACHE PATH "Root of LLVM installation")
9    set(LLVM_LIB ${LLVM_HOME}/lib)
10   set(LLVM_DIR ${LLVM_LIB}/cmake/llvm)
11   find_package(LLVM REQUIRED CONFIG)
12   include_directories(${LLVM_INCLUDE_DIRS})
13   link_directories(${LLVM_LIBRARY_DIRS})
14   set(SOURCE_FILE MethodRename.cpp)
15   add_executable(methodrename ${SOURCE_FILE})
16   set_target_properties(methodrename PROPERTIES COMPILE_FLAGS "-fno-rtti")
17   target_link_libraries(methodrename
18     LLVMSupport
19     clangAST
20     clangBasic
21     clangFrontend
22     clangSerialization
23     clangToolingCore
24     clangRewrite
25     clangTooling
26   )
27 endif()
```

Figure 7.8: Build configuration for 'methodrename' test tool

The most notable changes, compared to the code from Figure 3.20, are at *Lines 23 and 24,* where we added two new libraries to support code modifications: `clangToolingCore` and `clangRewrite` . Other changes include the new name for the tool (*Line 2*) and the source file that contains the main function (*Line 14*).

As soon as we finish with the code, it's time to build and run our tool.

7.2.6   Running the code modification tool

The program can be compiled using the same sequence of commands as we used previously in *Section 3.3, AST traversal*, see Figure 3.11:

```
export LLVM_HOME=<...>/llvm-project/install
mkdir build
cd build
cmake -G Ninja -DCMAKE_BUILD_TYPE=Debug ...
ninja
```

Figure 7.9: Configure and build commands for 'methodrename' tool

We can run the create tool on the following test file (`TestClass.cpp`):

```
 1 class TestClass {
 2 public:
 3   TestClass(){};
 4   void pos(){};
 5 };
 6
 7 int main() {
 8   TestClass test;
 9   test.pos();
10   return 0;
11 }
```

Figure 7.10: Original TestClass.cpp

We can run the tool as follows:

```
$ ./methodrename TestClass.cpp -- -std=c++17
Renamed method: pos to test_pos
Renamed method call: pos to test_pos
```

Figure 7.11: Running methodrename Clang Tool on TestClass.cpp

As we can see, the method `TestClass::pos` was renamed to `TestClass::test_pos`. The method call was also updated, as shown in the following figure:

```
1  class TestClass {
2  public:
3    TestClass(){};
4    void test_pos(){};
5  };
6
7  int main() {
8    TestClass test;
9    test.test_pos();
10    return 0;
11 }
```

Figure 7.12: Modified TestClass.cpp

The provided example demonstrates how Clang can assist in creating refactoring tools. The created Clang Tool uses a recursive visitor to set up the required code transformation. Another possible option is to use Clang-Tidy, which we investigated earlier in *Chapter 5, Clang-Tidy Linter Framework*. Let's examine this option in more detail.

7.3   Clang-Tidy as a code modification tool

We plan to investigate `FixItHint`, which is a part of the Clang Diagnostics subsystem (see *Section 4.4.2, Diagnostics support*). `FixItHint` can be integrated with `clang::Rewriter` and `clang::tooling::Replacement` explored previously, providing advanced diagnostics that are used in powerful tools such as Clang-Tidy.

7.3.1   FixItHint

`clang::FixItHint` is a class in the Clang compiler that significantly enhances its diagnostic capabilities. Its primary role is to provide automated suggestions for correcting code errors or issues that the compiler detects. These suggestions, known as "fix-its," are a part of Clang's diagnostic messages and are intended to guide developers in resolving identified issues in their code.

When Clang encounters a coding error, warning, or stylistic issue, it generates a `FixItHint`. This hint contains specific recommendations for changes in the source code. For instance, it may suggest replacing a snippet of text with a corrected version or inserting or removing code at a particular location.

For example, consider the following source code:

```
1 void foo() {
2   constexpr int a = 0;
3   constexpr const int *b = &a;
4 }
```

Figure 7.13: Test file foo.cpp

If we run a compilation for the file, we will get the following error:

```
$ <...>/llvm-project/install/bin/clang -cc1 -emit-obj foo.cpp -o /tmp/foo.o
foo.cpp:3:24: error: constexpr variable 'b' must be initialized by a
constant expression
    3 |    constexpr const int *b = &a;
      |                           ^   ~~
foo.cpp:3:24: note: pointer to 'a' is not a constant expression
foo.cpp:2:17: note: address of non-static constexpr variable 'a' may differ
on each invocation of the enclosing function; add 'static' to give it a
constant address
    2 |    constexpr int a = 0;
      |                  ^
      |    static
1 error generated.
```

Figure 7.14: Compilation error generated in foo.cpp

As you can see, the compiler suggests adding the **static** keyword at *Line 2* for the program shown in Figure 7.13.

The error is processed by Clang using the FixItHint object, as shown in Figure 7.15. As seen in Figure 7.15, when Clang detects an issue in the source code and generates a diagnostic, it can also produce a clang::FixItHint that suggests how to fix the issue. The hint is later processed by the Clang diagnostics subsystem and displayed to the user.

It's important to highlight that the hint can also be converted into a Replacement object, which represents the exact text change needed. For example, Clang-Tidy uses the Replacement object as temporary storage for information from FixItHint in its DiagnosticConsumer class implementation, allowing the FixItHint to be converted into a Replacement object that represents the exact text change needed.

```
if (VarD && VarD->isConstexpr()) {
  // Non-static local constexpr variables have unintuitive semantics:
  //   constexpr int a = 1;
  //   constexpr const int *p = &a;
  // ... is invalid because the address of 'a' is not constant.
  ↪ Suggest
  // adding a 'static' in this case.
  Info.Note(VarD->getLocation(), diag::note_constexpr_not_static)
      << VarD
      << FixItHint::CreateInsertion(VarD->getBeginLoc(), "static ");
```

Figure 7.15: Code fragment from clang/lib/AST/ExprConstant.cpp

Overall, `clang::FixItHint` enhances the user-friendliness and utility of Clang, providing developers with practical tools for improving code quality and resolving issues efficiently. Its integration into Clang's diagnostic system exemplifies the compiler's emphasis on not only pinpointing code issues but also aiding in their resolution. We are going to utilize this feature in a Clang-Tidy check that will rename methods in a test class and convert the code shown in Figure 7.10 to that in Figure 7.12.

7.3.2 Creating project skeleton

Let's create the project skeleton for our Clang-Tidy check. We will name our check "methodrename" and it will be a part of "misc" set of Clang-Tidy checks. We will use the command from Section 5.4.1

```
$ ./clang-tools-extra/clang-tidy/add_new_check.py misc methodrename
```

Figure 7.16: Creating a skeleton for the misc-methodrename check

The command from Figure 7.16 should be run from the root of the cloned LLVM project. We specified two parameters for the add_new_check.py script: misc – the set of checks that will contain our new check, and methodrename – the name of our check.

The command will produce the following output:

```
Updating ./clang-tools-extra/clang-tidy/misc/CMakeLists.txt...
Creating ./clang-tools-extra/clang-tidy/misc/MethodrenameCheck.h...
Creating ./clang-tools-extra/clang-tidy/misc/MethodrenameCheck.cpp...
Updating ./clang-tools-extra/clang-tidy/misc/MiscTidyModule.cpp...
Updating clang-tools-extra/docs/ReleaseNotes.rst...
Creating clang-tools-extra/test/clang-tidy/checkers/misc/methodrename.cpp...
Creating clang-tools-extra/docs/clang-tidy/checks/misc/methodrename.rst...
Updating clang-tools-extra/docs/clang-tidy/checks/list.rst...
Done. Now it's your turn!
```

Figure 7.17: Artefacts created for misc-methodrename check

We have to modify at least two generated files in the ./clang-tools-extra/clang-tidy/misc folder:

1. MethodrenameCheck.h : This is the header file for our check. Here, we want to add an additional private method processMethod for checking the method's properties and displaying diagnostics.

2. MethodrenameCheck.cpp : This file contains the processing logic, and we need to implement three methods: registerMatchers, check, and the newly added private method processMethod.

7.3.3   Check implementation

We will start with modifications to the header file:

```
27 private:
28   void processMethod(const clang::CXXMethodDecl *Method,
29                       clang::SourceLocation StartLoc, const char
                    ↪   *LogMessage);
30 };
```

Figure 7.18: MethodrenameCheck.h modifications

The added private method `MethodrenameCheck::processMethod` has the same parameters as the method introduced earlier in our Clang Tool 'methodrename', as seen in Figure 7.4.

We start the implementation with the `MethodrenameCheck::registerMatchers` method of our check as follows:

```
26 void MethodrenameCheck::registerMatchers(MatchFinder *Finder) {
27   auto ClassMatcher = hasAncestor(cxxRecordDecl(matchesName("::Test.*$")));
28   auto MethodMatcher = cxxMethodDecl(isNotTestMethod(), ClassMatcher);
29   auto CallMatcher = cxxMemberCallExpr(callee(MethodMatcher));
30   Finder->addMatcher(MethodMatcher.bind("method"), this);
31   Finder->addMatcher(CallMatcher.bind("call"), this);
32 }
```

Figure 7.19: Implementation of registerMatchers

Lines 30 and 31 register two matchers. The first one is for method declarations (bound to the "method" identifier), and the second one is for method calls (bound to the "call" identifier).

Here, we use a **Domain Specific Language (DSL)** defined in *Section 3.5, AST matchers*. The `ClassMatcher` specifies that our method declaration has to be declared within a class with a name starting with the "Test" prefix.

The method declaration matcher (MethodMatcher) is defined at *Line 28*. It must be declared within the class specified by ClassMatcher and should be a test method (details about the isNotTestMethod matcher will be described below).

The last matcher, CallMatcher, is defined at *Line 29* and specifies that it must be a call to a method that satisfies the conditions of MethodMatcher.

The isNotTestMethod matcher is an ad-hoc matcher that is used to check our specific conditions. We can define our own matchers using AST_MATCHER and related macros. The implementation for it can be found here:

```
18 AST_MATCHER(CXXMethodDecl, isNotTestMethod) {
19   if (Node.getAccess() != clang::AS_public) return false;
20   if (llvm::isa<clang::CXXConstructorDecl>(&Node)) return false;
21   if (!Node.getIdentifier() || Node.getName().startswith("test_")) return
     ↪ false;
22
23   return true;
24 }
```

Figure 7.20: isNotTestMethod matcher implementation

The macro has two parameters. The first one specifies the AST node we want to check, which is clang::CXXMethodDecl in our case. The second parameter is the matcher name that we want to use for the user-defined matcher, which is isNotTestMethod in our case.

The AST node can be accessed as a Node variable at the macro body. The macro should return true if the Node matches the required conditions. We use the same conditions we used for our 'methodrename' Clang Tool in Figure 7.4 (*Lines 46-51*).

The `MethodrenameCheck::check` is the main method for our check and can be implemented as follows:

```
34  void MethodrenameCheck::check(const MatchFinder::MatchResult &Result) {
35    if (const auto *Method = Result.Nodes.getNodeAs<CXXMethodDecl>("method"))
      ↪ {
36      processMethod(Method, Method->getLocation(), "Method");
37    }
38
39    if (const auto *Call = Result.Nodes.getNodeAs<CXXMemberCallExpr>("call"))
      ↪ {
40      if (CXXMethodDecl *Method = Call->getMethodDecl()) {
41        processMethod(Method, Call->getExprLoc(), "Method call");
42      }
43    }
44  }
```

Figure 7.21: check implementation

The code has two blocks. The first one (*Lines 35-37*) processes method declarations, and the last one (*Lines 39-42*) processes method calls. Both call `MethodrenameCheck::processMethod` to display diagnostics and create the required code modifications.

Let's examine how it's implemented and how clang::FixItHint is used.

```
46 void MethodrenameCheck::processMethod(const clang::CXXMethodDecl *Method,
47                                       clang::SourceLocation StartLoc,
48                                       const char *LogMessage) {
49   diag(StartLoc, "%0 %1 does not have 'test_' prefix") << LogMessage <<
     ↪  Method;
50   diag(StartLoc, "insert 'test_'", DiagnosticIDs::Note)
51       << FixItHint::CreateInsertion(StartLoc, "test_");
52 }
```

Figure 7.22: processMethod implementation

We print diagnostics about the detected issue at *Line 49*. *Lines 50-51* print an informational message about the suggested code modifications and create the corresponding code replacement at *Line 51*. To insert text, we use clang::FixItHint::CreateInsertion. We also display the insertion as a note for our primary warning.

As soon as all the required changes are applied to the generated skeleton, it's time to build and run our check on a test file.

7.3.4   Build and run the check

We assume that build configuration from Figure 1.12 was used. Thus, we have to run the following command to build our check:

```
$ ninja clang-tidy
```

We can install it to the install folder with:

```
$ ninja install
```

We can run our check as follows on the `TestClass` from Figure 7.10:

```
$ <...>/llvm-project/install/bin/clang-tidy \
          -checks='-*,misc-methodrename' \
          ./TestClass.cpp                  \
          -- -std=c++17
```

Figure 7.23: Clang-Tidy misc-methodrename check run on the test file TestClass.cpp

The command will produce the following output:

```
TestClass.cpp:4:8: warning: Method 'pos' does not have 'test_' prefix
[misc-methodrename]
  void pos(){};
       ^
TestClass.cpp:4:8: note: insert 'test_'
  void pos(){};
       ^
       test_
TestClass.cpp:9:8: warning: Method call 'pos' does not have 'test_' prefix
[misc-methodrename]
  test.pos();
       ^
TestClass.cpp:9:8: note: insert 'test_'
  test.pos();
       ^
       test_
```

Figure 7.24: Warning generated for TestClass.cpp by misc-methodrename check

As we can see, the check correctly detected two places where the method name has to be changed and created replacements. The command from Figure 7.23 does not modify the original source file. We have to specify an additional argument `-fix-notes` to apply the insertions specified as notes to the original warnings. The required command will look like this:

```
$ <...>/llvm-project/install/bin/clang-tidy \
            -fix-notes                      \
            -checks='-*,misc-methodrename' \
            ./TestClass.cpp                 \
            -- -std=c++17
```

Figure 7.25: Clang-Tidy with -fix-notes option

The command output is as follows:

```
2 warnings generated.
TestClass.cpp:4:8: warning: Method 'pos' does not have 'test_' prefix
[misc-methodrename]
  void pos(){};
       ^
TestClassSmall.cpp:4:8: note: FIX-IT applied suggested code changes
TestClass.cpp:4:8: note: insert 'test_'
  void pos(){};
       ^
       test_
TestClass.cpp:9:8: warning: Method call 'pos' does not have 'test_' prefix
[misc-methodrename]
  test.pos();
       ^
TestClass.cpp:9:8: note: FIX-IT applied suggested code changes
TestClass.cpp:9:8: note: insert 'test_'
```

```
test.pos();
       ^
   test_
clang-tidy applied 2 of 2 suggested fixes.
```

Figure 7.26: Clang-Tidy fixes applied to the TestClass.cpp

As we can see, the required insertions were applied here. Clang-Tidy has powerful tools to control the applied fixes and can be considered a significant resource for code modification. Another popular tool used for code modification is Clang-Format. As the name suggests, this tool specializes in code formatting. Let's explore it in detail.

7.4   Code modification and Clang-Format

Clang-Format is an essential tool in the Clang/LLVM project, designed for formatting C, C++, Java, JavaScript, Objective-C, or Protobuf code. It plays a crucial role in the Clang tooling ecosystem, offering capabilities for parsing, analyzing, and manipulating source code.

Clang-Format is a part of Clang and has to be installed if we have built and installed the Clang compiler. Let's look at how it can be used.

7.4.1   Clang-Format configuration and usage examples

Clang-Format uses .clang-format configuration files. The utility will use the closest configuration file; i.e., if the file is located at the folder with the source files we want to format, then the configuration from the folder will be used. The format for configuration files is YAML, which is the same format used for Clang-Tidy configuration files, as shown in *Section 5.3.2, Clang-Tidy configuration*. Let's create the following simple configuration file:

```
1 BasedOnStyle: LLVM
```

Figure 7.27: Simple .clang-format configuration file

The configuration file says that we will use the code style defined by LLVM, see `https:` `//llvm.org/docs/CodingStandards.html`.

Suppose we have a non-formatted file `main.cpp`, then the following command will format it:

```
$ <...>/llvm-project/install/bin/clang-format -i main.cpp
```

The result of the formatting is shown here:

```
 1  namespace clang {
 2  class TestClang {
 3  public:
 4  void testClang(){};
 5  };
 6  }int main() {
 7  TestClang test;
 8  test.testClang();
 9  return 0;
10  }
```

Original code

```
 1  namespace clang {
 2  class TestClang {
 3  public:
 4    void testClang(){};
 5  };
 6  } // namespace clang
 7  int main() {
 8    TestClang test;
 9    test.testClang();
10    return 0;
11  }
```

Formatted code

Figure 7.28: Formatting for main.cpp

In the example provided in Figure 7.28, we can see that the indentation defined by the LLVM code style was applied. We can also observe that Clang-Format broke *Line 6* in the original source code and made the main function definition start on a separate line. Additionally, we can see that Clang-Format added a comment to the namespace closing bracket in the formatted code at *Line 6*.

After considering the usage example, it's time to look at the internal design of Clang-Format.

7.4.2   Design considerations

At the core of Clang-Format is the Clang Lexer (see *Section 2.2.2, Lexer*), which tokenizes the input source code, breaking it down into individual tokens like keywords, identifiers, and literals. These tokens serve as the basis for formatting decisions.

The initial Clang-Format design document considered the Parser and AST as basic components for formatting. Despite the advantages provided by advanced data structures such as the AST, this approach has some disadvantages:

- The Parser requires a full build process and, therefore, build configuration.

- The Parser has limited capabilities to process a part of the source text, which is a typical task for formatting, such as formatting a single function or a source range of the source file.

- Formatting macros is a challenging task when using the AST as the basic structure for formatting. For instance, the processed macro may not be called in the compiled code and, as a result, may be missed in the AST.

- The Parser is much slower than the Lexer.

Clang-Format leverages `clang::tooling::Replacement` to represent code formatting changes and utilizes `clang::Rewriter` to apply these changes to the source code.

Configuration plays a pivotal role in Clang-Format's operation. Users define their preferred formatting style by configuring rules in a `.clang-format` file. This configuration specifies details such as indentation width, brace placement, line breaks, and more.

Clang-Format supports various predefined and customizable formatting styles, such as "LLVM," "Google," and "Chromium." Users can select a style that aligns with their project's coding standards.

Once tokenized, Clang-Format processes the token stream, taking into account the current context, indentation level, and configured style rules. It then adjusts whitespace and line breaks accordingly to adhere to the chosen style.

One notable feature of Clang-Format is its ability to handle macros effectively, preserving the original formatting within macros and complex macros.

Customization is a key aspect of Clang-Format. Users can extend or customize its behavior by defining custom rules and formatting options in the configuration file. This flexibility allows teams to enforce specific coding standards or adapt Clang-Format to project-specific needs.

It offers a user-friendly command-line interface, enabling manual code formatting or integration into scripts and automation.

Clang-Format utilizes Clang's Format library to generate formatted code accurately. This library ensures that the code consistently follows the desired formatting style. The design follows the main paradigm of LLVM: "everything is a library," as discussed in *Section 1.2.1, Short LLVM history*. Thus, we can effectively use the formatting functionality in other Clang Tools. For instance, formatting can be used with Clang-Tidy to format code with fixes applied by Clang-Tidy. Let's consider an example of how this functionality can be used.

7.4.3 Clang-Tidy and Clang-Format

The applied Clang-Tidy fixes can break formatting. Clang-Tidy suggests using the `-format-style` option to address the problem. This option will apply formatting using the functionality provided by the clangFormat library. The formatting is applied to the modified lines of code. Consider an example when our TestClass has broken formatting.

If we run Clang-Tidy as we did before (see Figure 7.25), then the formatting will remain unchanged and broken:

```
 1 class TestClass {
 2 public:
 3   TestClass(){};
 4 void pos(){};
 5 };
 6
 7 int main() {
 8   TestClass test;
 9 test.pos();
10   return 0;
11 }
```

```
 1 class TestClass {
 2 public:
 3   TestClass(){};
 4 void test_pos(){};
 5 };
 6
 7 int main() {
 8   TestClass test;
 9 test.test_pos();
10   return 0;
11 }
```

Original code   *Applied fixes*

Figure 7.29: Applying Clang-Tidy fixes without formatting on TestClassNotFormated.cpp

We used the following command for Figure 7.29

```
$ <...>/llvm-project/install/bin/clang-tidy  \
  -fix-notes                                 \
  -checks='-*,misc-methodrename'             \
  ./TestClassNotFormated.cpp                 \
  -- -std=c++17
```

The result will be different if we run Clang-Tidy with `-format-style` option, for example:

```
$ <...>/llvm-project/install/bin/clang-tidy  \
  -format-style 'llvm'                        \
  -fix-notes                                  \
  -checks='-*,misc-methodrename'              \
  ./TestClassNotFormated.cpp                  \
```

```
-- -std=c++17
```

As we can see the 'llvm' formatting style was chosen for the example. The result is shown in the following figure:

```
1 class TestClass {
2 public:
3   TestClass(){};
4 void pos(){};
5 };
6
7 int main() {
8   TestClass test;
9 test.pos();
10   return 0;
11 }
```

```
1 class TestClass {
2 public:
3   TestClass(){};
4   void test_pos(){};
5 };
6
7 int main() {
8   TestClass test;
9   test.test_pos();
10   return 0;
11 }
```

Original code   *Applied fixes with formatting*

Figure 7.30: Applying Clang-Tidy fixes with formatting on TestClassNotFormated.cpp

The relationship between Clang-Tidy and Clang-Format, as we just demonstrated, can be visualized as presented in the following figure:

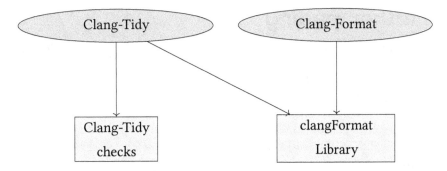

Figure 7.31: Clang-Tidy and Clang-Format integration

In the figure, both Clang-Tidy and Clang-Format use the `clangFormat` library to format the code.

The provided example demonstrates the integration of various Clang Tools. Modularity, an essential design decision in LLVM/Clang, is a key component for such integration. This example is not unique, and we will explore the further integration of different Clang Tools to enhance the development experience in **Integrated Development Environments (IDEs)** like **Visual Studio Code (VS Code)**. This will be the topic of our next chapter.

7.5 Summary

In this chapter, we investigated the different options provided by Clang for code modifications. We created a specialized Clang Tool that renames a method in a test class. We also rewrote the tool using Clang-Tidy and explored how custom AST matchers can be created. Furthermore, we delved into a variety of different classes provided by Clang for code modifications. One of these classes, `clang::FixItHint`, is integrated with the Clang diagnostics subsystem and provides a powerful tool for code modification within Clang, as well as in different tools created with Clang. We concluded with Clang-Format, the only tool in the book that does not use the AST but instead utilizes the Clang Lexer to perform code formatting. The next chapter will focus on the integration of different Clang Tools within IDEs.

7.6 Further reading

- Clang-Format Style Options: `https://clang.llvm.org/docs/ClangFormatStyleOptions.html`

- Peter Goldsborough, Emitting Diagnostics in Clang [23]

- AST Matcher Reference: `https://clang.llvm.org/docs/LibASTMatchersReference.html`

8

IDE Support and Clangd

This chapter is about the **Language Server Protocol (LSP)** and how you can utilize it to enhance your **Integrated Development Environment (IDE)**. Our primary IDE of choice is **Visual Studio Code (VS Code)**. LLVM has its own implementation of LSP known as **Clangd**. We will begin by describing LSP and exploring how Clangd leverages it to extend the capabilities provided by the IDE. Finally, we will conclude with examples of how various Clang tools, such as Clang-Tidy and Clang-Format, can be seamlessly integrated into the IDE through Clangd.

We will cover the following topics in this chapter:

- What is Language Server Protocol (LSP) and how does it improve an IDE's capabilities?

- How VS Code and Clangd (the Clang LSP server) can be installed

- How LSP is used to connect VS Code and Clangd, through an example

- How Clangd is integrated with other Clang tools

- Why performance matters for Clangd and what optimizations were made to make Clangd fast

8.1   Technical requirements

The source code for this chapter is located in the `chapter8` folder of the book's GitHub repository: `https://github.com/PacktPublishing/Clang-Compiler-Frontend-Packt/tree/main/chapter8`

8.2   Language Server Protocol

An IDE is a software application or platform that provides a comprehensive set of tools and features to assist developers in creating, editing, debugging, and managing software code. An IDE typically includes a code editor with syntax highlighting, debugging capabilities, project management features, version control integration, and, often, plugins or extensions to support various programming languages and frameworks.

Popular examples of IDEs are Visual Studio/VS Code, IntelliJ IDEA, Emacs, and Vim. These tools are designed to streamline the development process, making it easier for developers to write, test, and maintain their code efficiently.

A typical IDE supports multiple languages, and integrating each language can be a challenging task. Each language requires specific support, which can be visualized in Figure 8.1. It's worth noting that there are many similarities in the development process of different programming languages. For example, the languages shown in Figure 8.1 have a code navigation feature that allows developers to quickly locate and view the definition of a symbol or identifier within their code base.

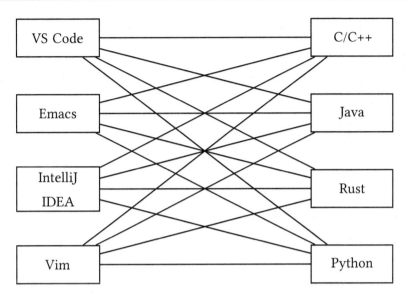

Figure 8.1: Programming languages integration in IDEs

The feature will be referred to as **go-to definition** in this chapter. Such similarities suggest a way to simplify the relationships shown in Figure 8.1 by introducing an intermediate level called the **Language Server Protocol**, or **LSP**, as shown here:

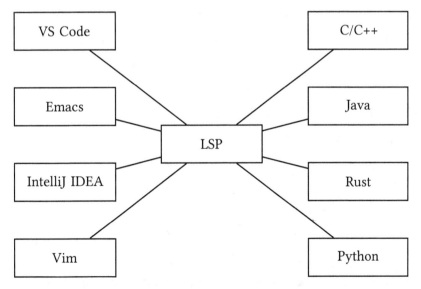

Figure 8.2: Programming languages integration in IDEs using LSP

The **LSP** project was initiated by Microsoft in 2015 as part of its efforts to improve VS Code, a lightweight, open source code editor. Microsoft recognized the need for a standardized way to provide rich language services across different programming languages within VS Code and other code editors.

LSP quickly gained popularity and adoption in the developer community. Many code editors and IDEs, including VS Code, Emacs, and Eclipse, began implementing support for LSP.

Language server implementations emerged for various programming languages. These language servers, developed by both Microsoft and the open source community, offered language-specific intelligence and services, making it easier to integrate language features into different editors.

In this chapter, we will explore **Clangd**, a language server that is part of clang-tools-extra. Clangd leverages the Clang compiler frontend and offers a comprehensive suite of code analysis and language support features. Clangd assists developers with intelligent code completion, semantic analysis, and real-time diagnostics, helping them to write code more efficiently and catch errors early in the development process. We will delve into Clangd in detail here, starting with a real example of the interaction between the IDE (VS Code) and Clangd. We will begin with the environment setup, including the Clangd build and VS Code setup.

8.3   Environment setup

We will begin our environment setup by building Clangd. Then, we will install VS Code, set up the Clangd extension, and configure Clangd within it.

8.3.1   Clangd build

It's worth building Clangd in release mode, as we did for LLDB in *Section 1.3.3, The LLVM debugger, its build, and usage*. This is because performance is crucial in IDEs. For instance, Clangd requires building an AST to provide code navigation functionality. If a user modifies a document, the document should be rebuilt, and the navigation functionality will not be available until the rebuild process is completed. This can result in delays in IDE responses. To prevent IDE slowness in responses, we should ensure that Clangd is built with all the required optimizations. You can use the following project configuration command:

```
cmake -G Ninja -DCMAKE_BUILD_TYPE=Release -DCMAKE_INSTALL_PREFIX=../install
 ↪  -DLLVM_TARGETS_TO_BUILD="X86"
 ↪  -DLLVM_ENABLE_PROJECTS="clang;clang-tools-extra" ../llvm
```

Figure 8.3: Release configuration for Clangd build

The command has to be run from the `release` folder that we created in *Section 1.3.3, The LLVM debugger, its build, and usage,* as shown in Figure 1.8. As you can see, we have enabled two projects in Figure 8.3: `clang` and `clang-tools-extra`.

You can use the following command to build and install Clangd:

```
$ ninja install-clangd -j $(nproc)
```

This command will utilize the maximum available threads on the system and install the binary into the folder specified in our CMake command in Figure 8.3, which is the `install` folder under the LLVM source tree.

After building the Clangd binary, our next step will include installing VS Code and configuring it to work with Clangd.

8.3.2 VS Code installation and setup

You can download and install VS Code from the VS Code website: `https://code.visuals tudio.com/download`.

The first step after running VS Code is to install the **Clangd** extension. An open source extension is available to work with Clangd via LSP. The extension's source code can be found on GitHub: `https://github.com/clangd/vscode-clangd`. However, we can easily install the latest version of the extension directly from within VS Code.

To do this, press *Ctrl+Shift+X* (or ⌘+*Shift+X* for macOS) to open the extensions panel. Search for Clangd and click the **Install** button.

Figure 8.4: Installing the Clangd extension

After installing the extension, we need to set it up. The main step is to specify the path to the Clangd executable.

You can access this setting via the **File | Preferences | Settings** menu or by pressing *Ctrl +* , (or ⌘+, for macOS), as shown in the following screenshot:

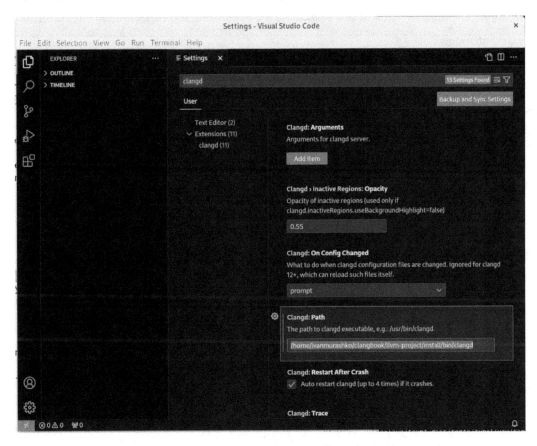

Figure 8.5: Setting up the Clangd extension

As shown in Figure 8.5, we have configured the Clangd path to be /home/ivanmurashko /clangbook/llvm-project/install/bin/clangd . This path was used during the installation of the Clangd binary in *Section 8.3.1, Clangd build.*

You can open your favorite C/C++ source file and try to navigate through it. For instance, you can search for a definition for a token, switch between a source and a header, and so on. In our next example, we will investigate how navigation, and especially go-to definition, works through LSP.

> **Important note**
>
> Our setup works only for simple projects that do not require special compilation flags. If your project requires special configuration to build, then you have to use a generated `compile_commands.json` file that should be placed at the root of your project. This file should contain a **Compilation Database (CDB)** in JSON format, specifying compilation flags for each file in your project. For more information about the setup, please refer to *Chapter 9, Clangd Setup for Large Projects*.

With the required components installed, we are now ready for an LSP demo where we will emulate typical development activities in an IDE (open and modify a document, jump to a token definition, etc.) and explore how it's represented via LSP.

8.4   LSP demo

In this brief LSP demo, we will demonstrate how Clangd opens a file and finds a symbol's definition. Clangd features a comprehensive logging subsystem that offers valuable insights into its interaction with the IDE. We will use the log subsystem to obtain the necessary information.

8.4.1   Demo description

In our example, we open a test file as shown in the following screenshot and retrieve the definition of the `doPrivateWork` token:

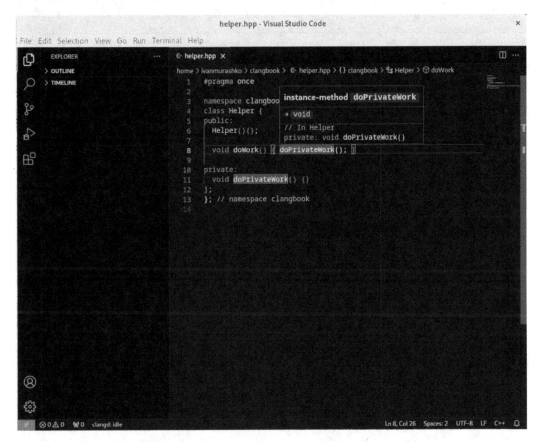

Figure 8.6: Go-to definition and hover for the `doPrivateWork` *token*

VS Code communicates with Clangd via standard input/output, and we will use Clangd logs to capture the interaction.

This can be achieved by setting up a wrapper shell script instead of using the actual clangd binary in the VS Code settings:

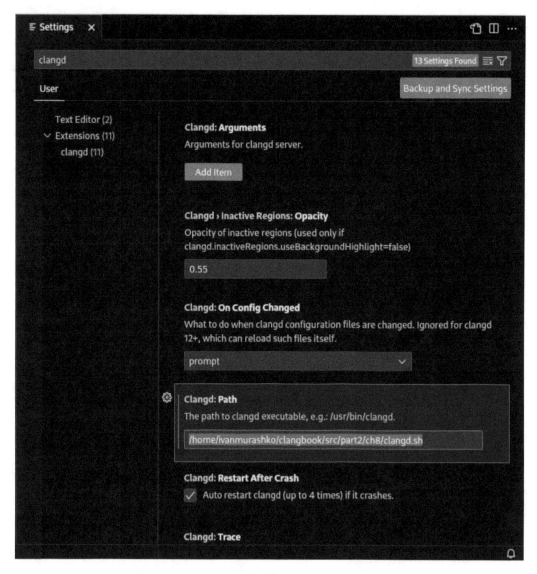

Figure 8.7: Wrapper shell script setup in VS Code

We can use the following script, `clangd.sh`:

```
1  #!/bin/sh
2  $HOME/clangbook/llvm-project/install/bin/clangd -log verbose -pretty 2>
   ↳   /tmp/clangd.log
```

Figure 8.8: Wrapper shell script for clangd

In Figure 8.8, we use two log options:

- The first one, `-log verbose`, activates verbose logging to ensure that actual LSP messages from and to Clangd will be logged.

- The second option, `-pretty`, is used to provide nicely formatted JSON messages. We also redirect stderr output to the log file, `/tmp/clangd.log`, in our case.

As a result, the file will contain logs from our example session. We can view these logs using the following command::

```
$ cat /tmp/clangd.log
```

In the logs, we can find "`textDocument/definition`" that was sent by VS Code:

```
V[16:24:39.336] <<< {
  "id": 13,
  "jsonrpc": "2.0",
  "method": "textDocument/definition",
  "params": {
    "position": {
      "character": 26,
      "line": 7
    },
```

```
    "textDocument": {
      "uri": "file:///home/ivanmurashko/clangbook/helper.hpp"
    }
  }
}
```

Figure 8.9: The "textDocument/definition" request sent by the IDE

The request sent by the IDE is received and processed by Clangd. The corresponding log is recorded as follows:

```
I[16:24:39.336] <-- textDocument/definition(13)
V[16:24:39.336] ASTWorker running Definitions on version 1 of /home/.../
helper.hpp
```

Figure 8.10: Handling of the "textDocument/definition" request in Clangd

Finally, Clangd creates the response and sends it to the IDE. The corresponding log record shows that the reply was sent:

```
I[16:24:39.336] --> reply:textDocument/definition(13) 0 ms
V[16:24:39.336] >>> {
  "id": 13,
  "jsonrpc": "2.0",
  "result": [
    {
      "range": {
        "end": {
          "character": 20,
          "line": 10
        },
```

```
      "start": {
        "character": 7,
        "line": 10
      }
    },
    "uri": "file:///home/ivanmurashko/clangbook/helper.hpp"
    }
  ]
}
```

Figure 8.11: The "textDocument/definition" reply from Clangd

The logs will be our primary tool to investigate LSP internals. Let's dive into more complex examples.

8.4.2 LSP session

An LSP session consists of several requests to and responses from the Clangd server. It starts with an "initialize" request. Then, we open a document, and VS Code sends a "textDocument/didOpen" notification. After the request, Clangd will periodically respond with "textDocument/publishDiagnostics" notifications when the state of the opened file changes. For example, this occurs when compilation is finished and its ready to process navigation requests. Next, we initiate a go-to definition request for a token, and Clangd responds with the location information for the found definition. We also investigate how Clangd processes file modifications that are notified by the client via "textDocument /didChange" notifications. We finish our session with a "textDocument/didClose" request when we close the opened file. A diagram depicting the interaction is presented in the following figure:

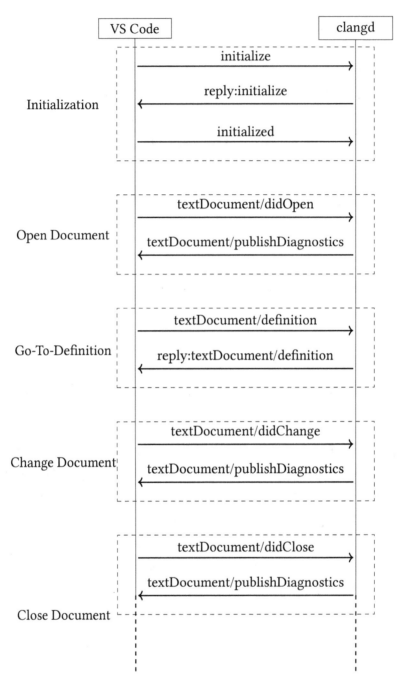

Figure 8.12: LSP session example

Let's look at the example in detail. We will start with the "initialize" request.

Initialization

To establish communication, the client (code editor or IDE) and the language server exchange JSON-RPC messages. The initialization process begins with the client sending an "initialize" request to the language server, specifying the capabilities it supports. The actual request sent by VS Code is quite large, and a simplified version, where some parts of the request are replaced with "...", is shown as follows:

```
1  {
2    "id": 0,
3    "jsonrpc": "2.0",
4    "method": "initialize",
5    "params": {
6      "capabilities": {
7        ...
8        "textDocument": {
9          ...
10          "definition": {
11            "dynamicRegistration": true,
12            "linkSupport": true
13          },
14          ...
15        },
16        "clientInfo": {
17          "name": "Visual Studio Code",
18          "version": "1.85.1"
19        },
20        ...
21      }
22  }
```

Figure 8.13: VS Code to Clangd (initialize request)

In the request, the client (VS Code) tells the server (Clangd) what capabilities are supported on the client side; for example, at *Lines 10-13* in Figure 8.13, the client says that it supports the "textDocument/definition" request type that is used for go-to definition requests.

The language server replies to the request with a response that contains capabilities supported by the server:

```
 1 {
 2   "id": 0,
 3   "jsonrpc": "2.0",
 4   "result": {
 5     "capabilities": {
 6       ...
 7       "definitionProvider": true,
 8       ...
 9     },
10     "serverInfo": {
11       "name": "clangd",
12       "version": "clangd version 16.0.6
          ↪ (https://github.com/llvm/llvm-project.git
          ↪ 7cbf1a2591520c2491aa35339f227775f4d3adf6) linux
          ↪ x86_64-unknown-linux-gnu"
13     }
14   }
15 }
```

Figure 8.14: Clangd to VS Code (initialize reply)

As we can see, the same id is used to connect the request with its reply. Clangd replied that it supports go-to definition requests as specified in *Line 7* in Figure 8.14. Thus our client (VS Code) can send the navigation request to the server, which we will explore later in *Section 8.4.2, Go-to definition.*

VS Code acknowledges the initialization by sending an "initialized" notification:

```
1 {
2   "jsonrpc": "2.0",
3   "method": "initialized"
4 }
```

Contrary to the "initialize" request, there is a notification, and it does not expect any response from the server. As a result, it does not have an "id" field. The "initialized" notification can be sent only once, and it should be received before any other requests or notifications are sent from the client side. After the initialization, we are ready to open a document and send the corresponding "textDocument/didOpen" notification.

Open document

When a developer opens a C++ source file, the client sends a "textDocument/didOpen" notification to inform the language server about the newly opened file. In our example, the opened file is located at /home/ivanmurashko/clangbook/helper.hpp , and the corresponding notification sent by VS Code will look like this:

```
{
  "jsonrpc": "2.0",
  "method": "textDocument/didOpen",
  "params": {
    "textDocument": {
      "languageId": "cpp",
      "text": "#pragma once\n\nnamespace clangbook {\nclass Helper
        ↪ {\npublic:\n  Helper(){};\n\n  void doWork() { doPrivateWork();
        ↪ }\n\nprivate:\n  void doPrivateWork() {}\n};\n}; // namespace
        ↪ clangbook\n",
      "uri": "file:///home/ivanmurashko/clangbook/helper.hpp",
```

```
      "version": 1
    }
  }
}
```

Figure 8.15: VS Code to Clangd (didOpen notification)

As we can see, VS Code sends the notification with parameters included in the "params/ textDocument" field. These parameters consist of the filename in the "uri" field and the source file text within the "text" field.

Clangd starts compiling the file upon receiving the 'didOpen' notification. It builds an AST and extracts semantic information about different tokens from it. The server uses this information to distinguish between different tokens with the same name. For example, we can use a token named 'foo' that may serve as a class member or a local variable depending on the scope in which it is used, as shown in the following code fragment:

```
1 class TestClass {
2 public:
3   int foo(){return 0};
4 };
5
6 int main() {
7   TestClass test;
8   int foo = test.foo();
9   return foo;
10 }
```

Figure 8.16: Occurrences of the 'foo' token in foo.hpp

As we can see in *Line 8*, we use the `'foo'` token two times: as a function call and in a local variable definition.

The go-to definition request will be delayed until the compilation process is finished. It's worth noting that the majority of requests are put in a queue and wait until the compilation process is finished. The rule has some exemptions, and some requests can be executed without an AST with a limited provided functionality. One of the examples is the code-formatting requests. The code formatting does not require an AST and therefore the formatting functionality can be provided before the AST is built.

If the state of the file is changed, then Clangd will notify VS Code with the `"textDocument/publishDiagnostics"` notification. For example, when the compilation process is finished, then Clangd will send the notification to VS Code:

```
1 {
2   "jsonrpc": "2.0",
3   "method": "textDocument/publishDiagnostics",
4   "params": {
5     "diagnostics": [],
6     "uri": "file:///home/ivanmurashko/clangbook/helper.hpp",
7     "version": 1
8   }
9 }
```

Figure 8.17: Clangd to VS Code (publishDiagnostics notification)

As we can see, there are no compilation errors; `params/diagnostics` is empty. It will contain errors or warning descriptions if our code contains a compilation error or warning, as shown here:

```
 1 {
 2   "jsonrpc": "2.0",
 3   "method": "textDocument/publishDiagnostics",
 4   "params": {
 5     "diagnostics": [
 6       {
 7         "code": "expected_semi_after_expr",
 8         "message": "Expected ';' after expression (fix available)",
 9         "range": {
10           "end": {
11             "character": 35,
12             "line": 7
13           },
14           "start": {
15             "character": 34,
16             "line": 7
17           }
18         },
19         "relatedInformation": [],
20         "severity": 1,
21         "source": "clang"
22       }
23     ],
24     "uri": "file:///home/ivanmurashko/clangbook/helper.hpp",
25     "version": 5
26   }
27 }
```

Figure 8.18: Clangd to VS Code (publishDiagnostics with compilation error)

VS Code processes the diagnostics and displays it, as shown in the following screenshot:

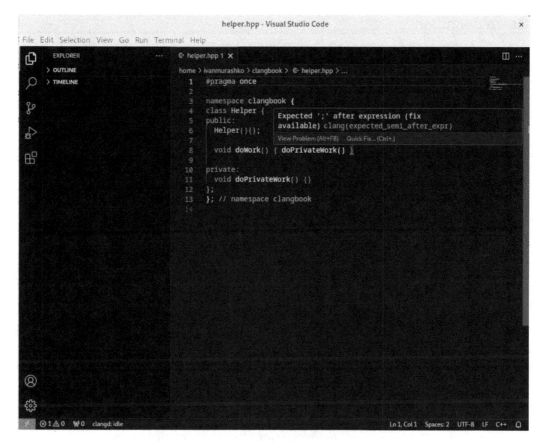

Figure 8.19: Compilation error in helper.hpp

After the compilation finished and we got "textDocument/publishDiagnostics", Clangd is ready to process navigation requests, such as "textDocument/definition" (go-to definition).

Go-to definition

To find the definition of a symbol in a C++ file, the client sends a "textDocument/definition" request to the language server:

```
 1 {
 2   "id": 13,
 3   "jsonrpc": "2.0",
 4   "method": "textDocument/definition",
 5   "params": {
 6     "position": {
 7       "character": 26,
 8       "line": 7
 9     },
10     "textDocument": {
11       "uri": "file:///home/ivanmurashko/clangbook/helper.hpp"
12     }
13   }
14 }
```

Figure 8.20: VS Code to Clangd (textDocument/definition request)

The line position is specified as 7 instead of the actual line 8 in the editor, as shown in Figure 8.6. This is because line numbering starts at 0.

The language server responds with the definition location in the C++ code:

```
 1  {
 2    "id": 13,
 3    "jsonrpc": "2.0",
 4    "result": [
 5      {
 6        "range": {
 7          "end": {
 8            "character": 20,
 9            "line": 10
10          },
11          "start": {
12            "character": 7,
13            "line": 10
14          }
15        },
16        "uri": "file:///home/ivanmurashko/clangbook/helper.hpp"
17      }
18    ]
19  }
```

Figure 8.21: Clangd to VS Code (textDocument/definition response)

As we can see, the server responded with the actual position of the definition. Another popular action in the IDE is document modification. This functionality is served by the "textDocument/didChange" notification. Let's look at it.

Change document

As part of the document modification, let's insert a comment, `// Constructor`, at *Line 6*, as shown in this screenshot:

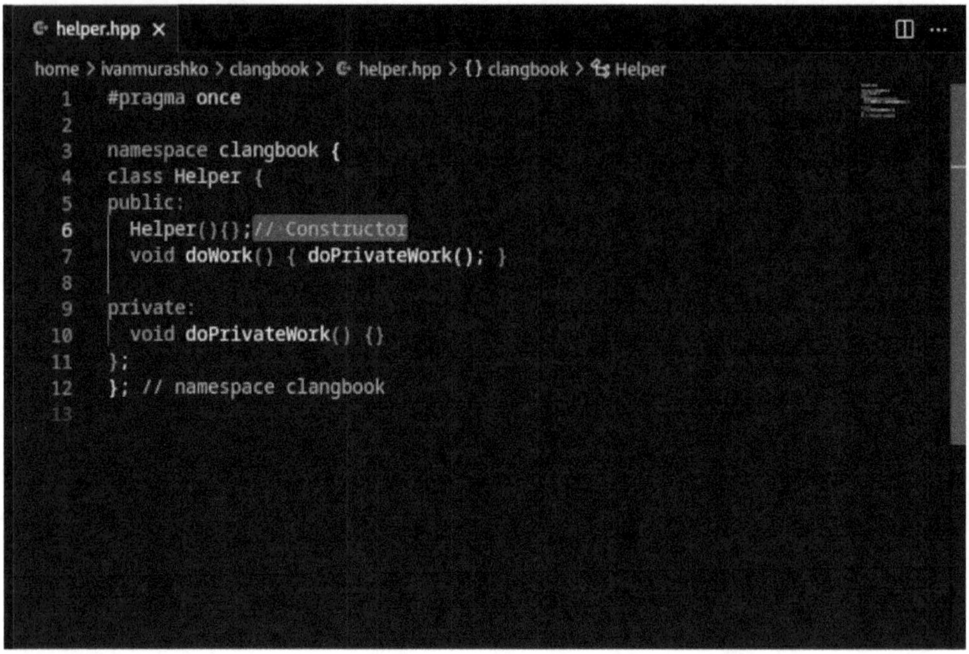

Figure 8.22: Change document

VS Code will detect that the document has been modified and notify the LSP server (Clangd) using the following notification:

```
 1 {
 2   "jsonrpc": "2.0",
 3   "method": "textDocument/didChange",
 4   "params": {
 5     "contentChanges": [
 6       {
 7         "range": {
 8           "end": {
 9             "character": 13,
10             "line": 5
11           },
12           "start": {
13             "character": 13,
14             "line": 5
15           }
16         },
17         "rangeLength": 0,
18         "text": "// Constructor"
19       }
20     ],
21     "textDocument": {
22       "uri": "file:///home/ivanmurashko/clangbook/helper.hpp",
23       "version": 2
24     }
25   }
26 }
```

Figure 8.23: VS Code to Clangd (didChange notification)

As we can see, the notification contains the range specification and the text for replacing the specified range in the document. One important part of the notification is the "version" field, which specifies the version of the document.

We can observe that version changed from 1, as used in the document open (see *Line 9* in Figure 8.15), to 2 for the document modification (see *Line 23* in Figure 8.23).

Clangd starts the document compilation because the document modification can cause a significant change in the resulting AST, which is used for navigation requests. Once the compilation is finished, the server will respond with the corresponding "textDocument/publishDiagnostics" notification, as shown here:

```
1 {
2     "jsonrpc": "2.0",
3     "method": "textDocument/publishDiagnostics",
4     "params": {
5         "diagnostics": [],
6         "uri": "file:///home/ivanmurashko/clangbook/helper.hpp",
7         "version": 2
8 }
```

Figure 8.24: Clangd to VS Code (publishDiagnostics notification)

As we can see, the diagnostic was sent for the modified document because it contains the version field pointing to version 2, which corresponds to the modified document, as seen in *Line 7* of Figure 8.24.

Our last action in the example is to close the document. Let's take a closer look at it.

Closing a document

When we finish our work with the document and close it, VS Code sends a
"textDocument/didClose" notification to the language server:

```
1 {
2   "jsonrpc": "2.0",
3   "method": "textDocument/didClose",
4   "params": {
5     "textDocument": {
6       "uri": "file:///home/ivanmurashko/clangbook/helper.hpp"
7     }
8   }
9 }
```

Figure 8.25: VS Code to Clangd (textDocument/didClose request)

After receiving the request, Clangd will remove the document from its internal structures.
Clangd will not send any updates for the document anymore, thus it will empty out the list
of diagnostics shown on the client (e.g., in the **Problems** pane of VS Code) by sending the
final empty "textDocument/publishDiagnostics" message, as shown here:

```
1 {
2   "jsonrpc": "2.0",
3   "method": "textDocument/publishDiagnostics",
4   "params": {
5     "diagnostics": [],
6     "uri": "file:///home/ivanmurashko/clangbook/helper.hpp"
7   }
8 }
```

Figure 8.26: Clangd to VS Code (textDocument/didClose request)

The shown example demonstrates the typical interactions between Clangd and VS Code.

The provided example utilizes functionality from the Clang frontend, that is, basic Clang functionality. On the other hand, Clangd has a strong connection with other Clang tools, such as Clang-Format and Clang-Tidy, and can reuse the functionality provided by these tools. Let's take a closer look at this.

8.5   Integration with Clang tools

Clangd takes advantage of the LLVM module architecture and has a very strong integration with other Clang tools. In particular, Clangd uses Clang-Format libraries to provide formatting functionality and Clang-Tidy libraries (such as libraries with clang-tidy checks) to support linters in the IDE. The integration is schematically shown in the following figure:

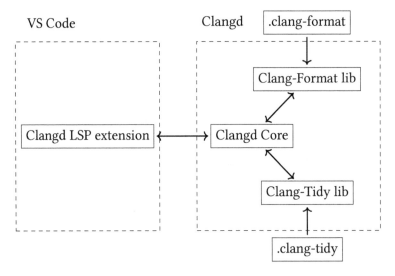

Figure 8.27: VS Code with LSP extension and Clangd server for C++

The configuration from .clang-format (see *Section 7.4.1, Clang-Format configuration and usage examples*) is used for formatting, and from .clang-tidy (see *Section 5.3.2, Clang-Tidy configuration*) for linters. Let's see how the formatting works in Clangd.

8.5.1 Clangd support for code formatting using LSP messages

Clangd provides robust support for code formatting. This feature is essential for developers to maintain consistent code styles and readability in their C and C++ projects. Clangd leverages LSP messages, primarily the "textDocument/formatting" and "textDocument/rangeFormatting" requests, to achieve this functionality.

Formatting entire documents

The "textDocument/formatting" request is used when a developer wants to format the entire content of a document. This request is typically initiated by the user in VS Code by pressing *Ctrl + Shift + I* (or ⌘+ *Shift + I* for macOS); the IDE sends a "textDocument/formatting" request to Clangd for the entire document:

```
1  {
2    "id": 9,
3    "jsonrpc": "2.0",
4    "method": "textDocument/formatting",
5    "params": {
6      "options": {
7        "insertSpaces": true,
8        "tabSize": 4
9      },
10     "textDocument": {
11       "uri": "file:///home/ivanmurashko/clangbook/helper.hpp"
12     }
13   }
14 }
```

Figure 8.28: VS Code to Clangd (textDocument/formatting request)

Clangd processes this request by utilizing the code style configuration specified in the project's .clang-format file. The .clang-format file contains formatting rules and

preferences, allowing developers to define their desired code style; see *Section 7.4.1,*
Clang-Format configuration and usage examples.

The response contains the list of modifications to be applied to the opened document:

```
1  {
2    "id": 9,
3    "jsonrpc": "2.0",
4    "result": [
5      {
6        "newText": "\n    ",
7        "range": {
8          "end": {
9            "character": 0,
10           "line": 5
11         },
12         "start": {
13           "character": 7,
14           "line": 4
15         }
16       }
17     }
18   ]
19 }
```

Figure 8.29: Clangd to VS Code (textDocument/formatting response)

In the example, we should replace the text at the specified range at *Lines 7-16* in Figure 8.29
with new text specified at *Line 6*.

Formatting specific code ranges

In addition to formatting entire documents, Clangd also supports formatting specific code
ranges within a document. This is achieved using the "textDocument/rangeFormatting"

request. Developers can select a range within the code, such as a function, a block of code, or even just a few lines, and request formatting for that specific range, as shown in the following screenshot:

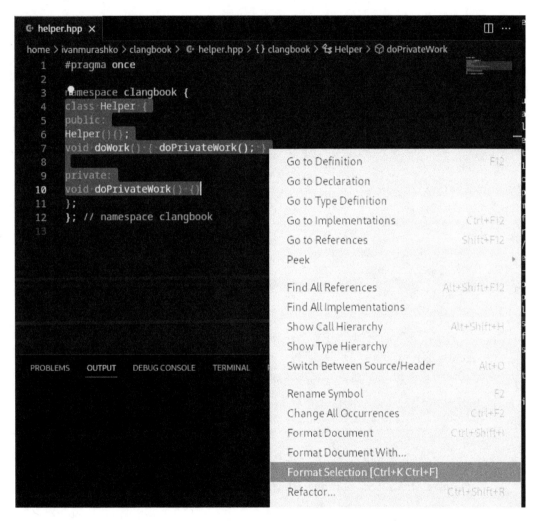

Figure 8.30: Reformatting a specific code range in helper.hpp

When selecting the menu item or pressing *Ctrl + K* and then *Ctrl + F* (or ⌘+ *K* and then ⌘+ *F* for macOS), VS Code will send the following request to Clangd:

```
 1 {
 2    "id": 89,
 3    "jsonrpc": "2.0",
 4    "method": "textDocument/rangeFormatting",
 5    "params": {
 6      "options": {
 7        "insertSpaces": true,
 8        "tabSize": 4
 9      },
10      "range": {
11        "end": {
12          "character": 2,
13          "line": 10
14        },
15        "start": {
16          "character": 0,
17          "line": 3
18        }
19      },
20      "textDocument": {
21        "uri": "file:///home/ivanmurashko/clangbook/helper.hpp"
22      }
23    }
24 }
```

Figure 8.31: VS Code to Clangd (textDocument/rangeFormatting request)

The "`textDocument/rangeFormatting`" request specifies the range to be formatted within the document, and Clangd applies the same formatting rules from the `.clang-format` file to this specific code segment. The response will be similar to the one used for the formatting request and will contain the modification that should be applied to the original text, as shown in Figure 8.29. The only difference will be the method name, which should be "textDocument/rangeFormatting" in this case.

Another tool that is integrated via Clangd is Clang-Tidy, and it utilizes the LSP protocol in a different manner compared to the formatting functionality that we just described.

8.5.2 Clang-Tidy

As we can see, Clangd uses specific LSP methods to implement integration with Clang-Format:

- "textDocument/formatting"

- "textDocument/rangeFormatting"

On the other hand, the integration with Clang-Tidy is implemented differently, and it reuses the "publishDiagnostics" notification to report linter warnings and errors.

Let's investigate how it works and create a custom Clang-Tidy configuration as the first step.

Clang-Tidy integration with LSP

We will run the misc-methodrename check that we recently created for testing method renaming, see *Section 7.3, Clang-Tidy as a code modification tool.* Our Clang-Tidy configuration will look like this:

```
1 Checks: '-*,misc-methodrename'
```

Figure 8.32: .clang-tidy config for IDE integration

The .clang-tidy file with the configuration should be placed in the folder with our test project.

If we rename our helper class to TestHelper, we will be able to observe that the lint check we created in *Section 7.3, Clang-Tidy as a code modification tool* will start reporting diagnostics about the incorrect method name used for the test class. The corresponding diagnostic is displayed in the drop-down pane and in the **PROBLEMS** tab, as we can see in the following screenshot:

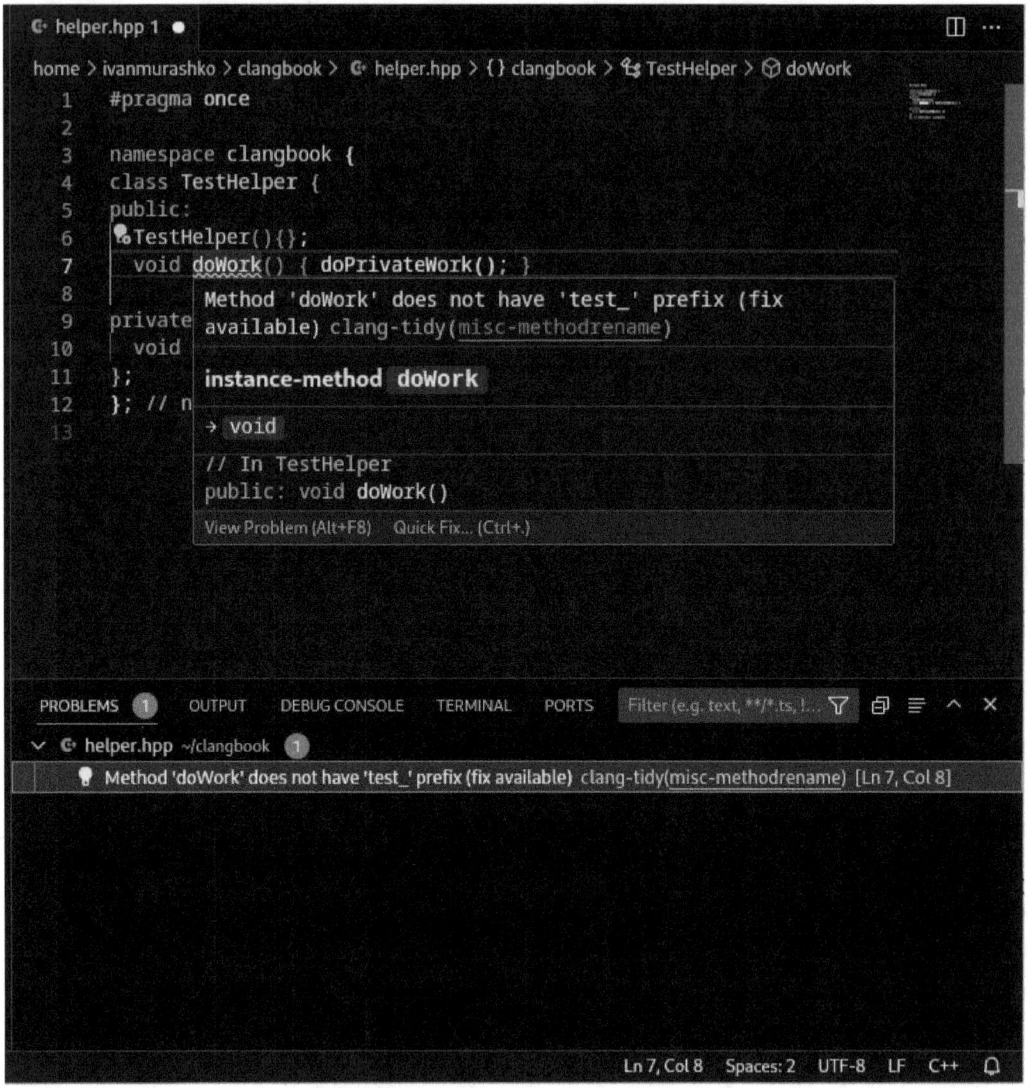

Figure 8.33: Clang-Tidy integration

The message is displayed as part of diagnostics. Specifically, the following notification is sent from Clang to VS Code:

```
1  {
2    "jsonrpc": "2.0",
3    "method": "textDocument/publishDiagnostics",
4    "params": {
5      "diagnostics": [
6        {
7          "code": "misc-methodrename",
8          "codeDescription": {
9            "href": "https://clang.llvm.org/extra/clang-tidy/checks/misc/
10                     methodrename.html"
11         },
12         "message": "Method 'testdoWork' does not have 'test_' prefix (fix
   ↪ available)",
13         "range": {
14           "end": {
15             "character": 17,
16             "line": 6
17           },
18           "start": {
19             "character": 7,
20             "line": 6
21           }
22         },
23         "relatedInformation": [],
24         "severity": 2,
25         "source": "clang-tidy"
26       }
27     ],
28     "uri": "file:///home/ivanmurashko/clangbook/helper.hpp",
29     "version": 11
30   }
```

Figure 8.34: Clangd to VS Code (publishDiagnostics notification)

As we can see in the figure (*Line 11*), a fix for the problem is also available. There is an amazing opportunity to apply Clang-Tidy fixes in the IDE. Let's explore how the functionality is implemented with LSP.

Applying fixes in the IDE

The fixes can be applied in the IDE and the functionality is provided via the "textDocument/codeAction" method. The method is used by VS Code to prompt Clangd to compute commands for the specific document and range. The most important parts of the command are provided in the following example:

```
 1  {
 2    "id": 98,
 3    "jsonrpc": "2.0",
 4    "method": "textDocument/codeAction",
 5    "params": {
 6      "context": {
 7        "diagnostics": [
 8          {
 9            "code": "misc-methodrename",
10            ...
11            "range": ...,
12          ...
13        },
14        ...
15      }
16  }
```

Figure 8.35: VS Code to Clangd (textDocument/codeAction request)

The most important part of the request is at *Lines 7-11*, where we can see a copy of the original diagnostics notification. This information will be used to retrieve the necessary document modifications provided by clang::FixItHint in the activated check.

Consequently, Clangd can respond with the action that describes the required modification to be made:

```
1  {
2    "id": 98,
3    "jsonrpc": "2.0",
4    "result": [
5      {
6        "diagnostics": [
7          ...
8        ],
9        "edit": {
10         "changes": {
11           "file:///home/ivanmurashko/clangbook/helper.hpp": [
12             {
13               "newText": "test_",
14               "range": {
15                 "end": {
16                   "character": 7,
17                   "line": 6
18                 },
19                 "start": {
20                   "character": 7,
21                   "line": 6
22                 }
23               }
24             }
25             ...
26         }
27      }
28  }
```

Figure 8.36: Clangd to VS Code (codeAction response)

The "edit" field in Figure 8.36 is the most important part of the response, as it describes the changes to be applied to the original text.

The integration with Clang-Tidy is possible without extra computation because the AST is built for navigation and diagnostics purposes by Clangd core. The AST can be used as the seed for Clang-Tidy checks, eliminating the need to run a separate Clang-Tidy executable to retrieve messages from the linter. This is not the only optimization made in Clangd; let's now look at another example of performance optimizations in Clangd.

8.6   Performance optimizations

Obtaining a smooth IDE experience with accurate results provided without visible delays is a challenging task. One of the ways to achieve this experience is through compiler performance optimization, as good navigation can be provided with well-parsed source code. Clangd offers excellent examples of performance optimization, which we will explore in some detail. We will start with the optimizations for code modifications.

8.6.1   Optimizations for modified documents

As we saw in *Section 8.4.2, Open document*, navigation support requires the AST as the basic data structure, so we have to use the Clang frontend to obtain it. Additionally, we have to rebuild the AST when there are document modifications. Document modification is a common activity for developers, and we won't be able to provide a good IDE experience if we always start the build process from scratch.

Source code preamble

To gain insights into the ideas used to speed up AST building for modified documents, let's examine a simple C++ program:

```
1 #include <iostream>
2
3 int main() {
4   std::cout << "Hello world!" << std::endl;
5   return 0;
6 }
```

Figure 8.37: C++ program: helloworld.cpp

The program has six lines of code, but the conclusion can be deceptive. The #include directive inserts a lot of additional code. We can estimate the amount of code inserted by the preprocessor if we run Clang with the -E command-line option and calculate the number of lines, as follows:

```
$ <...>/llvm-project/install/bin/clang -E helloworld.cpp | wc -l
36215
```

Figure 8.38: Number of lines in the post-processed program

where <...> is the folder where llvm-project was cloned; see Figure 1.1.

As we can see, the code that should be parsed contains more than 36,000 lines of code. This is a common pattern, and the majority of the code to be compiled is inserted from included headers. The part of the document located at the beginning of the source file and containing the include directives is called the **preamble**.

It's worth noting that preamble modifications are possible but rare, for instance, when we insert a new header. The majority of the modifications are located in the code outside the preamble.

The primary idea for performance optimization is to cache the preamble AST and reuse it for any compilation of a modified document.

AST build at Clangd

The performance optimization made in Clangd involves a two-part compilation process. In the first part, the preamble that contains all included header files is compiled into a precompiled header; see *Chapter 10, Precompiled headers*. This precompiled header is then used in the second stage of the compilation process to build the AST.

This complex process serves as a performance optimization, especially when a user makes changes to a file that requires recompilation. Although a significant portion of compilation time is spent on header files, these files are typically not modified frequently. To address this, Clangd caches the AST for header files within the precompiled header file.

As a result, when modifications are made outside header files, Clangd does not need to rebuild them from scratch. Instead, it can reuse the cached AST for headers, significantly improving compilation performance and reducing the time needed for recompilation when working with header files. If a user modification affects header files, then the entire AST should be rebuilt, resulting in a cache miss in such cases. It's worth noting that modifications to headers are not as common as modifications to the primary source code (outside the included headers). Therefore, we can expect a pretty good cache hit rate for ordinary document modifications.

The precompiled header can be stored on a disk as a temporary file but can also reside in memory, which can also be considered a performance optimization.

The cached preamble is a powerful tool that significantly improves Clangd's performance in processing document changes made by a user. On the other hand, we should always consider edge cases that involve preamble modification. The preamble can be modified in two main ways:

1. **Explicitly**: When the user explicitly modifies the preamble, for instance, by inserting a new header into it or deleting an existing one

2. **Implicitly**: When the user implicitly modifies the preamble, for instance, by modifying the headers that are included in the preamble

The first one can be easily detected via a "textDocument/didChange" notification that affects the range where the preamble is located. The second one is tricky, and Clangd should monitor the modifications in the included headers to correctly process navigation requests.

Clangd also has some modifications aimed at making preamble compilation faster. Some of these modifications required specific processing in Clang. Let's delve into it in detail.

8.6.2 Building preamble optimization

An interesting optimization can be applied to function bodies. A function body can be considered an essential part of primary indexing because it contains symbols that a user can click on, such as getting a definition for the symbol. This primarily applies to function bodies that are visible to the user in the IDE. On the other hand, many functions and their implementations (bodies) are hidden from the user in included headers. As a result, the user cannot request information about symbols from such function bodies. However, these bodies are visible to the compiler because it resolves include directives and parses the header files from the directives. The time spent by the compiler can be significant, considering that a complex project can have numerous dependencies, resulting in many header files being included in the document opened by the user. One obvious optimization is to skip function bodies when parsing header files from the preamble. This can be achieved using a special frontend option:

```
/// FrontendOptions - Options for controlling the behavior of the frontend.
class FrontendOptions {
  ...
  /// Skip over function bodies to speed up parsing in cases where you do
  ↪ not need
  /// them (e.g., with code completion).
  unsigned SkipFunctionBodies : 1;
```

```
    . . .
};
```

Figure 8.39: The FrontendOptions class from clang/Frontend/FrontendOptions.h

Clangd utilizes this option when building the preamble in the following manner:

```
 1 std::shared_ptr<const PreambleData>
 2 buildPreamble(PathRef FileName, CompilerInvocation CI,
 3               const ParseInputs &Inputs, bool StoreInMemory,
 4               PreambleParsedCallback PreambleCallback,
 5               PreambleBuildStats *Stats) {
 6   ...
 7   // Skip function bodies when building the preamble to speed up building
 8   // the preamble and make it smaller.
 9   assert(!CI.getFrontendOpts().SkipFunctionBodies);
10   CI.getFrontendOpts().SkipFunctionBodies = true;
11   ...
12   auto BuiltPreamble = PrecompiledPreamble::Build(...);
13   ...
14   // When building the AST for the main file, we do want the function
15   // bodies.
16   CI.getFrontendOpts().SkipFunctionBodies = false;
17   ...
18 };
```

Figure 8.40: buildPreamble from clang-tools-extra/clangd/Preamble.cpp

As we can see, Clangd uses the frontend option to skip function bodies in headers but disables it just before building the AST for the main document; see *Lines 10 and 16* in Figure 8.40.

Such optimization can significantly improve the document readiness time (when the opened document is ready for navigation requests from the user) for complex C++ source files.

While the performance optimizations discussed here offer valuable insights into Clangd's efficiency, it's important to note that Clangd employs a multitude of other techniques to ensure its reliability and speed. Clangd serves as an excellent platform for experimenting with and implementing various optimization strategies, making it a versatile environment for performance enhancements and innovations.

8.7 Summary

In this chapter, we acquired knowledge of LSP, a protocol used to provide developer tools integration with IDEs. We explored Clangd, an LSP server that is part of LLVM and can be considered as a prime example of how it integrates various tools discussed in the book. Clangd utilizes the Clang frontend to display compilation errors and leverages the AST as a fundamental data structure that provides information for navigation requests, such as go-to definition requests. Additionally, Clangd is seamlessly integrated with other tools covered in previous chapters, such as Clang-Tidy and Clang-Format. This integration showcases the significant benefits of the LLVM/Clang module structure.

8.8 Further reading

- Language Server Protocol specification: `https://microsoft.github.io/language-server-protocol/`

- Clangd documentation: `https://clangd.llvm.org/`

Part 3

Appendix

The Clang compiler is a very complex topic, and some details were omitted in the primary chapters of the book. The appendix covers some important aspects of Clang and Clang Tools that might be valuable if you are starting to apply this knowledge to complex projects containing many files and intricate build rules.

We will begin by discussing how Clang Tools can be integrated into a large project that uses complex compilation flags. LLVM serves as one of the examples of such projects.

Another important aspect is performance in complex C++ projects. Clang offers techniques that can be used to improve build speed for such projects, and we will also explore these features.

This part has the following chapters:

- *Chapter 9, Compilation Database*

- *Chapter 10, Build Speed Optimizations*

9

Appendix 1: Compilation Database

The test examples considered in the book do not require special compilation flags and typically can be compiled without any flags. However, this is not the scenario if you want to employ the material on a real project, such as running a lint check on your code base. In that situation, you will need to furnish special compilation flags for each file to be processed. Clang offers various methods for supplying these flags. We will explore in detail the JSON Compilation Database, which is one of the primary tools for delivering compilation flags to Clang tools such as Clang-Tidy and Clangd.

Compilation database definition

A **compilation database (CDB)** is a JSON file that specifies how each source file in a code base should be compiled. This JSON file is typically named `compile_commands.json` and resides in the root directory of a project. It provides a machine-readable record of all compiler invocations in the build process and is often used by various tools for more accurate analysis, refactoring, and more. Each entry in this JSON file typically contains the following fields:

- **directory**: The working directory of the compilation.

- **command**: The actual compile command, including compiler options.

- **arguments**: Another field that can be used to specify compilation arguments. It contains the list of arguments.

- **file**: The path to the source file being compiled.

- **output**: The path to the output created by this compilation step.

As we can see from the fields description, there are two ways to specify compilation flags: using the **command** or **arguments** field. Let's look at a specific example. Suppose our C++ file `ProjectLib.cpp` is located at the `/home/user/project/src/lib` folder and can be compiled with Clang using the following invocation command (the command is used as an example, and you can ignore the meaning of its arguments)

```
$ cd /home/user/project/src/lib
$ clang -Wall -I../headers ProjectLib.cpp -o ProjectLib.o
```

The following CDB can be used to represent the command:

```
1 [
2     {
3         "directory": "/home/user/project/src/lib",
4         "command": "clang -Wall -I../headers ProjectLib.cpp -o
          ↳  ProjectLib.o",
5         "file": "ProjectLib.cpp",
6         "output": "ProjectLib.o"
7     }
8 ]
```

Figure 9.1: Compilation Database for ProjectLib.cpp

The "command" field was used in the example. We can also create the CDB in another form and use the arguments field. The result will be as follows:

```
 1  [
 2      {
 3          "directory": "/home/user/project/src/lib",
 4          "arguments": [
 5              "clang",
 6              "-Wall",
 7              "-I../headers",
 8              "ProjectLib.cpp",
 9              "-o",
10              "ProjectLib.o"
11          ],
12          "file": "ProjectLib.cpp",
13          "output": "ProjectLib.o"
14      }
15  ]
```

Figure 9.2: CDB for ProjectLib.cpp

The **CDB** shown in Figure 9.2 represents the same compilation recipe as in Figure 9.1, but it uses a list of arguments (the "arguments" field) instead of the invocation command (the "command" field) used in Figure 9.1. It's important to note that the list of arguments also contains the executable "clang" as its first argument. CDB processing tools can use this argument to make a decision about which compiler should be used for the compilation in environments where different compilers are available, such as GCC versus Clang.

The provided CDB example contains only one record for one file. A real project might contain thousands of records. LLVM is a good example, and if you look at the build folder that we used for the LLVM build (see *Section 1.3.1, Configuration with CMake*), you may notice that it contains a compile_commands.json file with the CDB for the projects we selected to be built. It's worth noting that LLVM creates the CDB by default, but your project might require some special manipulations to create it. Let's look at how the CDB can be created in detail.

CDB creation

The `compile_commands.json` file can be generated in various ways. For example, the build system CMake has built-in support for generating a compilation database. Some tools can also generate this file from Makefiles or other build systems. There are even tools such as Bear and intercept-build that can generate a CDB by intercepting the actual compile commands as they are run.

So while the term is commonly associated with Clang and LLVM-based tools, the concept itself is more general and could theoretically be used by any tool that needs to understand the compilation settings for a set of source files. We will start with CDB generation using CMake, one of the most popular build systems.

Generating a CDB with CMake

Generating a CDB with CMake involves a few steps:

1. First, open a terminal or command prompt and navigate to your project's root directory.

2. Then, run CMake with the `-DCMAKE_EXPORT_COMPILE_COMMANDS=ON` option, which instructs CMake to create a `compile_commands.json` file. This file contains the compilation commands for all source files in your project.

3. After configuring your project with CMake, you can find the `compile_commands.json` file in the same directory where you ran the configuration command.

As we noticed before, LLVM created the CDB by default. It's achievable because `llvm/CMakeLists.txt` contains the following setup:

```
# Generate a CompilationDatabase (compile_commands.json file) for our
  ↪  build,
# for use by clang_complete, YouCompleteMe, etc.
set(CMAKE_EXPORT_COMPILE_COMMANDS 1)
```

Figure 9.3: LLVM-18.x CMake configuration from llvm/CMakeLists.txt

i.e., it set up the CDB generation by default.

Ninja to Generate a CDB

The Ninja can also be used to generate a CDB. We can use a Ninja subtool called "compdb" to dump the CDB to stdout. To run the subtool, we use the -t <subtool> command-line option in Ninja. Thus, we will use the following command to produce the CDB with Ninja:

```
$ ninja -t compdb > compile_commands.json
```

Figure 9.4: Creating a CDB with Ninja

This command instructs Ninja to generate the CDB information and save it in the compile_commands.json file.

The generated compilation database can be used with the different Clang tools that we have described in the book. Let's look at two of the most valuable examples, which include Clang-Tidy and Clangd.

Clang tools and a CDB

The concept of a CDB is not specific to Clang but Clang-based tools make extensive use of it. For instance, the Clang compiler itself can use a compilation database to understand how to compile files in a project. Tools such as Clang-Tidy and Clangd (for language support in IDEs) can also use it to ensure they understand code as it was built, making their analyses and transformations more accurate.

Clang-Tidy Configuration for Large Projects

To use clang-tidy with a CDB, you typically don't need any additional configuration. Clang-tidy can automatically detect and utilize the `compile_commands.json` file in your project's root directory.

On the other hand, Clang Tools provide a special option, **-p**, defined as follows:

```
-p <build-path> is used to read a compile command database
```

You can use this option to run Clang-Tidy on a file from the Clang source code. For example, if you run it from the llvm-project folder where the source code was cloned, it would look like this:

```
$ ./install/bin/clang-tidy clang/lib/Parse/Parser.cpp -p ./build/
```

Figure 9.5: Running Clang-Tidy on the LLVM code base

In this case, we are running Clang-Tidy from the folder, where we installed it, as described in *Section 5.2.1, Building and testing Clang-Tidy*. We have also specified the `build` folder as the project root folder containing the CDB.

Clang-Tidy is one of the tools that actively uses the CDB to be executed on large projects. Another tool is Clangd, which we will also explore.

Clangd Setup for Large Projects

Clangd offers a special configuration option to specify the path to the CDB. This option is defined as follows:

```
$ clangd --help

...

--compile-commands-dir=<string> - Specify a path to look for
compile_commands.json.If the path is invalid, clangd will search
```

```
in the current directory and parent paths of each source file.
```

...

Figure 9.6: Description for '–compile-commands-dir' option from 'clangd –help' output

You can specify this option in Visual Studio Code via the **Settings** panel, as shown in the following figure:

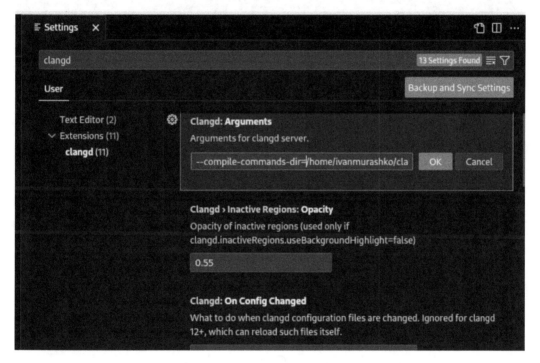

Figure 9.7: Configure the CDB path for clangd

Therefore, if you open a file from the Clang source code, you will have access to navigation support provided by Clangd as you can see in the following figure:

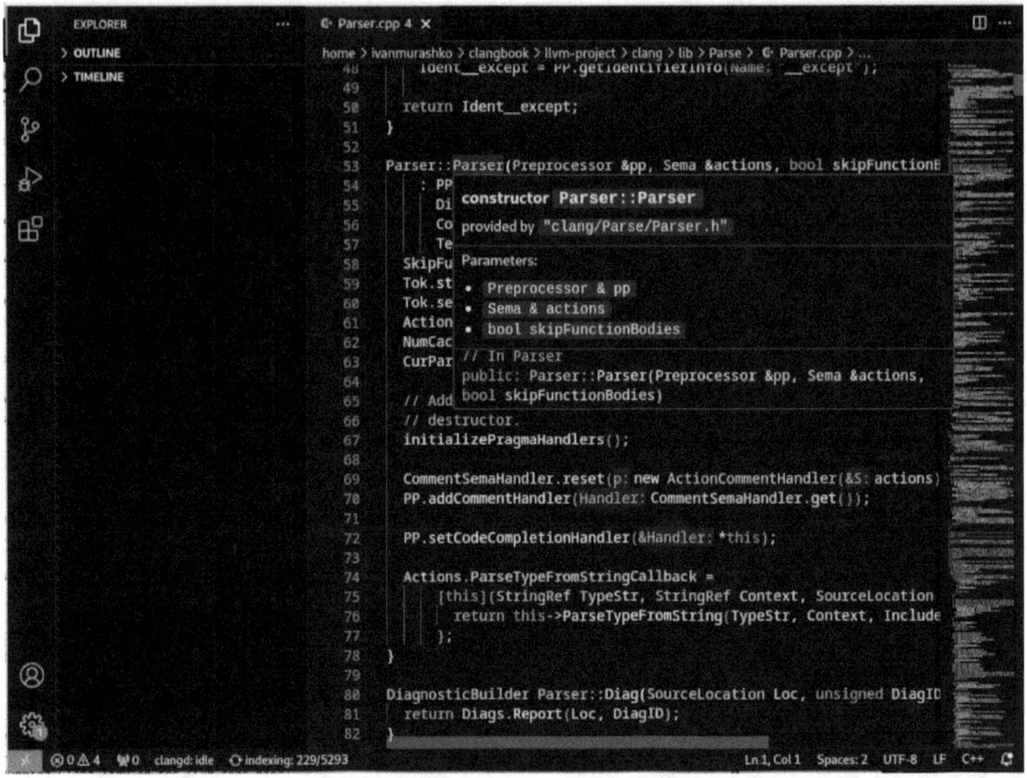

Figure 9.8: Hover provided for Parser::Parser method by Clangd at clang/lib/Parse/Parser.cpp

Integration of compile commands with Clang tools, such as Clang-Tidy or Clangd, provides a powerful tool for exploring and analyzing your source code.

Further reading

- Clang Documentation - JSON Compilation Database Format Specification: `https://clang.llvm.org/docs/JSONCompilationDatabase.html`

- Clangd documentation - Compile commands: `https://clangd.llvm.org/design/compile-commands`

10

Appendix 2: Build Speed Optimization

Clang has implemented several features with the goal of improving build speed for large projects. One of the most interesting features is precompiled headers and modules. They can be considered techniques that allow caching some parts of the AST and reusing it for different compiler invocations. Caching can significantly improve build speed for your project, and some of these features can be used to speed up different Clang tool executions. For instance, precompiled headers are used as the primary Clangd optimization for document editing.

In this appendix, we will cover two primary topics

- Precompiled headers

- Modules

Technical requirements

The source code for this appendix is located in the chapter10 folder of the book's GitHub repository: https://github.com/PacktPublishing/Clang-Compiler-Frontend-Packt /tree/main/chapter10.

Precompiled headers

Precompiled headers PCH, are a Clang feature designed to improve Clang's frontend performance. The basic idea is to create an AST (Abstract Syntax Tree) for a header file and reuse this AST during compilation for sources that include the header file.

Generating a precompiled header file is simple [5]. Suppose you have the following header file, header.h :

```
1 #pragma once
2
3 void foo() {
4 }
```

Figure 10.1: Header file to be compiled to PCH

You can generate a PCH for it with the following command:

```
$ <...>/llvm-project/install/bin/clang -cc1 -emit-pch            \
                                       -x c++-header header.h \
                                       -o header.pch
```

Here, we use the -x c++-header option to specify that the header file should be treated as a C++ header file. The output file will be named header.pch .

Simply generating precompiled headers is not enough; you need to start using them. A typical C++ source file that includes the header may look like this:

```
1 #include "header.h"
2
3 int main() {
4     foo();
5     return 0;
6 }
```

Figure 10.2: Source file that includes header.h

As you can see, the header is included as follows:

```
1 #include "header.h"
2
```

Figure 10.3: Header header.h inclusion

By default, Clang will not use a PCH, and you have to specify it explicitly with the following command:

```
$ <...>/llvm-project/install/bin/clang -cc1 -emit-obj           \
                                       -include-pch header.pch \
                                       main.cpp -o main.o
```

Here, we use -include-pch to specify the included precompiled header: header.pch .

You can check this command with a debugger, and it will give you the following output:

```
1 $ lldb <...>/llvm-project/install/bin/clang -- -cc1 -emit-obj -include-pch
  ↪  header.pch main.cpp -o main.o
2 ...
3 (lldb) b clang::ASTReader::ReadAST
4 ...
5 (lldb) r
6 ...
7 -> 4431    llvm::TimeTraceScope scope("ReadAST", FileName);
8    4432
9    4433    llvm::SaveAndRestore SetCurImportLocRAII(CurrentImportLoc,
     ↪  ImportLoc);
10   4434    llvm::SaveAndRestore<std::optional<ModuleKind>>
     ↪  SetCurModuleKindRAII(
11 (lldb) p FileName
12 (llvm::StringRef)  (Data = "header.pch", Length = 10)
```

Figure 10.4: Loading precompiled header at clang::ASTReader::ReadAST

From this example, you can see that Clang reads the AST from the precompiled header file. It's important to note that the precompiled header is read before parsing, allowing Clang to obtain all symbols from the header file before parsing the main source file. This makes explicit header inclusion unnecessary. Therefore, you can remove the #include "header.h" directive from the source file and achieve successful compilation.

This is impossible without precompiled headers, where you would encounter the following compilation error:

```
main.cpp:4:3: error: use of undeclared identifier 'foo'
    4 |   foo();
      |   ^
1 error generated.
```

Figure 10.5: Compilation error generated due to missing includes

It's worth noting that only the first -include-pch option will be processed; all others will be ignored. This reflects the fact that there can be only one precompiled header for a translation unit. On the other hand, a precompiled header can include another precompiled header. This functionality is known as chained precompiled headers [3], as it creates a chain of dependencies where one precompiled header depends on another precompiled header.

The usage of precompiled headers is not limited to regular compilation. As we saw in *Section 8.6.1, AST build at Clangd*, precompiled headers are actively used for performance optimizations in Clangd as placeholders for a cache for the preamble that contains included headers.

Precompiled headers are a technique that has been used for a long time, but they have some limitations. One of the most important limitations is that there can be only one precompiled header, which significantly limits the usage of PCH in real projects. Modules address some of the problems related to precompiled header. Lets explore them.

Clang modules

Modules, or **Precompiled Modules (PCMs)**, can be considered the next step in the evolution of precompiled headers. They also represent a parsed AST in binary form but form a DAG (tree), meaning one module can include more than one other module.

This is a major improvement compared to precompiled headers, where only one precompiled header can be introduced for each compilation unit.

The C++20 standard [21] introduced two concepts related to modules. The first one is ordinary modules, described in *section 10* of [21]. The other one is the so-called header unit , mostly described in *section 15.5*. Header units can be considered an intermediate step between ordinary headers and modules and allow the use of the import directive to import ordinary headers.

We will focus on Clang modules, which can be considered an implementation of header units from the C++ standard. There are two different options to use Clang modules. The first one is called **explicit modules**. The second is called **implicit modules**. We will explore both cases but will start with a description of a test project for which we want to use the modules.

Test project description

For experiments with modules, we will consider an example with two header files: header1.h and header2.h , which define the **void** foo1() and **void** foo2() functions, respectively, as shown:

```
1  #pragma once

2

3  void foo1() {}
```

```
1  #pragma once

2

3  void foo2() {}
```

          Header file: header1.h   *Header file: header2.h*

Figure 10.6: Header files to be used for the tests

These header files will be used in the following source file:

```cpp
1 #include "header1.h"
2 #include "header2.h"
3
4 int main() {
5   foo1();
6   foo2();
7   return 0;
8 }
```

Figure 10.7: Source file: main.cpp

We are going to organize our header files into modules. Clang uses a special file that contains the logical structure, which is called a **modulemap** file. Let's see what the file looks like for our test project.

Modulemap file

The modulemap file for our project will be named `module.modulemap` and has the following content:

```
1 module header1 {
2   header "header1.h"
3   export *
4 }
5 module header2 {
6   header "header2.h"
7   export *
8 }
```

Figure 10.8: Modulemap file: module.modulemap

As shown in Figure 10.8, we have defined two modules, **header1** and **header2**.

Each of them contains only one header and exports all symbols from it.

Now that we have collected all the necessary parts, we are ready to build and use the modules. Modules can be built explicitly or implicitly. Let's start with explicit builds.

Explicit modules

The module's structure is described by the modulemap file, as seen in Figure 10.8. Each of our modules has only one header, but a real module might include several headers. Thus, to build a module, we have to specify the structure of the modules (the modulemap file) and the module name we want to build. For instance, for the **header1** module, we can use the following build command:

```
$ <...>/llvm-project/install/bin/clang -cc1              \
       -emit-module -o header1.pcm                        \
       -fmodules module.modulemap -fmodule-name=header1 \
       -x c++-header -fno-implicit-modules
```

There are several important aspects in the compile command. The first one is the **-cc1** option, which indicates that we are calling only the compiler frontend. For more information, please refer to *Section 2.3, Clang driver overview*. Additionally, we specify that we want to create a build artifact (module) named header1.pcm by using the following option: -emit-module -o header1.pcm. The logical structure and the required modules to be built are specified in the module.modulemap file, which has to be specified as a compile argument with the -fmodule-name=header1 option. Enabling the modules functionality is done using the -fmodules flag, and we also specify that our headers are C++ headers with the -x c++-header option. To explicitly disable implicit modules, we include -fno-implicit-modules in the command because implicit modules, which we will investigate later in *Chapter 10, Implicit modules*, are enabled by default, but we don't want to use them at the moment.

The second module (header2) has a similar compilation command:

```
$ <...>/llvm-project/install/bin/clang -cc1              \
        -emit-module -o header2.pcm                      \
        -fmodules module.modulemap -fmodule-name=header2 \
        -x c++-header -fno-implicit-modules
```

The next step is to compile main.cpp using the generated modules, which can be done as follows:

```
$ <...>/llvm-project/install/bin/clang -cc1             \
        -emit-obj main.cpp                               \
        -fmodules -fmodule-map-file=module.modulemap \
        -fmodule-file=header1=header1.pcm                \
        -fmodule-file=header2=header2.pcm                \
        -o main.o -fno-implicit-modules
```

As we can see, both the module name and build artifacts (PCM files) are specified using the -fmodule-file compile option. The format used, such as header1=header1.pcm , indicates that header1.pcm corresponds to the header1 module. We also specify the modulemap file with the -fmodule-map-file option. It's worth noting that we created two build artifacts: header1.pcm and header2.pcm , and used them together for the compilation. This is impossible in the case of precompiled headers because only one precompiled header is allowed, as mentioned in *Chapter 10, Precompiled headers.*

We emitted an object file, main.o , as a result of the compilation command. The object file can be linked as follows:

```
$ <...>/llvm-project/install/bin/clang main.o -o main -lstdc++
```

Let's verify that the modules were loaded during compilation. This can be done with LLDB as follows:

```
1 $ lldb <...>/llvm-project/install/bin/clang -- -cc1 -emit-obj main.cpp
  ↪ -fmodules -fmodule-map-file=module.modulemap
  ↪ -fmodule-file=header1=header1.pcm -fmodule-file=header2=header2.pcm -o
  ↪ main.o -fno-implicit-modules
2 ...
3 (lldb) b clang::CompilerInstance::findOrCompileModuleAndReadAST
4 ...
5 (lldb) r
6 ...
7 Process 135446 stopped
8 * thread #1, name = 'clang', stop reason = breakpoint 1.1
9     frame #0: ... findOrCompileModuleAndReadAST(..., ModuleName=(Data =
      ↪ "header1", Length = 7), ...
10 ...
11 (lldb) c
12 Process 135446 stopped
13 * thread #1, name = 'clang', stop reason = breakpoint 1.1
14     frame #0: ... findOrCompileModuleAndReadAST(..., ModuleName=(Data =
      ↪ "header2", Length = 7), ....
15 ...
16 (lldb) c
17 Process 135446 resumed
18 Process 135446 exited with status = 0 (0x00000000)
```

Figure 10.9: Explicit module load

We set a breakpoint at clang::CompilerInstance::findOrCompileModuleAndReadAST, as shown in *Line 3* of Figure 10.9. We hit the breakpoint twice: first at *Line 9* for the module named header1 , and then at *Line 14* for the module named header2 .

You must explicitly define the build artifacts and specify the path where they will be stored in all compile commands when using explicit modules, as we have just discovered. However, all the required information is stored within the modulemap file (refer to Figure 10.8). The compiler can utilize this information to create all the necessary build artifacts automatically. The answer to the question is affirmative, and this functionality is provided by implicit modules. Let's explore it.

Implicit modules

As mentioned earlier, the modulemap file contains all the information required to build all modules (header1 and header2) and use them for dependent file (main.cpp) building. Thus, we have to specify a path to the modulemap file and a folder where the build artifacts will be stored. This can be done as follows:

```
$ <...>/llvm-project/install/bin/clang -cc1 \
    -emit-obj main.cpp                       \
    -fmodules                                \
    -fmodule-map-file=module.modulemap       \
    -fmodules-cache-path=./cache             \
    -o main.o
```

As we can see, we didn't specify -fno-implicit-modules , and we also specified the path for build artifacts with -fmodules-cache-path=./cache . If we examine the path, we will be able to see the created modules:

```
$ tree ./cache
./cache
|-- 2AL78TH69W6HR
    |-- header1-R65CPR1VCRM1.pcm
    |-- header2-R65CPR1VCRM1.pcm
    |-- modules.idx
2 directories, 3 files
```

Figure 10.10: The cache generated by Clang for implicit modules

Clang will monitor the cache folder (./cache in our case) and delete build artifacts that have not been used for a long time. It will also rebuild the modules if their dependencies (for instance, included headers) have changed.

Modules are a very powerful tool, but like every powerful tool, they can introduce non-trivial problems. Let's explore the most interesting problem that can be caused by modules.

Some problems related to modules

The code that uses modules can introduce some non-trivial behavior into your program. Consider a project that consists of two headers, as shown:

```
1 #pragma once
2
3 int h1 = 1;
```

```
1 #pragma once
2
3 int h2 = 2;
```

Header file: header1.h   *Header file: header2.h*

Figure 10.11: Header files that will be used for the test

The only header1.h is included in main.cpp, as follows

```
1 #include "header1.h"
2
3 int main() {
4     int h = h1 + h2;
5     return 0;
6 }
```

Figure 10.12: Source file: main.cpp

The code will not compile:

```
$ <...>/llvm-project/install/bin/clang  main.cpp -o main -lstdc++
main.cpp:4:16: error: use of undeclared identifier 'h2'
  int h = h1 + h2;
               ^
1 error generated.
```

Figure 10.13: Compilation error generated due to a missing header file

The error is obvious because we didn't include the second header that contains a definition for the h2 variable.

The situation would be different if we were using implicit modules. Consider the following module.modulemap file:

```
1 module h1 {
2   header "header1.h"
3   export *
4   module h2 {
5     header "header2.h"
6     export *
7   }
8 }
```

Figure 10.14: Modulemap file that introduces implicit dependencies

This file creates two modules, h1 and h2 . The second module is included within the first one.

If we compile it as follows, the compilation will be successful:

```
$ <...>/llvm-project/install/bin/clang -cc1 \
       -emit-obj main.cpp                   \
       -fmodules                            \
       -fmodule-map-file=module.modulemap   \
       -fmodules-cache-path=./cache         \
       -o main.o
$ <...>/llvm-project/install/bin/clang main.o -o main -lstdc++
```

Figure 10.15: Successful compilation for a file with a missing header but with implicit modules enabled

The compilation completed without any errors because the modulemap implicitly added header2.h to the used module (h1). We also exported all symbols using the export * directive. Thus, when Clang encounters #include "header1.h", it loads the corresponding h1 module, and therefore implicitly loads symbols defined in the h2 module and header2.h header.

The example illustrates how the visibility scope can be leaked when modules are used in the project. This can lead to unexpected behavior for the project build, when it builds with modules enabled and disabled.

Further reading

- Clang modules: https://clang.llvm.org/docs/Modules.html

- Precompiled header and modules internals: https://clang.llvm.org/docs/PCHInternals.html

Bibliography

[1] Alfred V. Aho, Monica S. Lam, Ravi Sethi, and Jeffrey D. Ullman. *Compilers: Principles, Techniques, and Tools*. Addison-Wesley, 2 edition, 2006. ISBN 978-0-321-48681-3.

[2] Bruno Cardoso Lopes and Nathan Lanza. [RFC] An MLIR based Clang IR (CIR). June 2022. URL `https://discourse.llvm.org/t/rfc-an-mlir-based-clang-ir-cir/63319`.

[3] LLVM Community. Precompiled Header and Modules Internals. URL `https://clang.llvm.org/docs/PCHInternals.html`.

[4] LLVM Community. Moving LLVM Projects to GitHub. 2019. URL `https://llvm.org/docs/Proposals/GitHubMove.html`.

[5] LLVM Community. Clang Compiler User's Manual. 2022. URL `https://clang.llvm.org/docs/UsersManual.html`.

[6] LLVM Community. [LLVM] Update C++ standard to 17. 2022. URL `https://reviews.llvm.org/D130689`.

[7] LLVM Community. Building LLVM with CMake. 2023. URL `https://llvm.org/docs/CMake.html`.

[8] LLVM Community. "Clang" CFE Internals Manual. 2023. URL `https://clang.llvm.org/docs/InternalsManual.html`.

[9] LLVM Community. MSVC compatibility. 2023. URL `https://clang.llvm.org/docs/MSVCCompatibility.html`.

[10] LLVM Community. Clang features. 2023. URL https://clang.llvm.org/features.html.

[11] LLVM Community. LLVM Coding Standards. 2023. URL https://llvm.org/docs/CodingStandards.html.

[12] LLVM Community. CommandLine 2.0 Library Manual. 2023. URL https://llvm.org/docs/CommandLine.html.

[13] LLVM Community. LLVM Programmer's Manual. 2023. URL https://llvm.org/docs/ProgrammersManual.html.

[14] LLVM Community. How to set up LLVM-style RTTI for your class hierarchy. 2023. URL https://llvm.org/docs/HowToSetUpLLVMStyleRTTI.html.

[15] LLVM Community. How To Build On ARM. 2024. URL https://llvm.org/docs/HowToBuildOnARM.html.

[16] LLVM Community. AST Matcher Reference. 2024. URL https://clang.llvm.org/docs/LibASTMatchersReference.html.

[17] LLVM Community. Extra Clang Tools documentation: Clang-Tidy. 2024. URL https://clang.llvm.org/extra/clang-tidy/.

[18] Keith Cooper and Linda Torczon. *Engineering A Compiler*. Elsevier Inc., 2nd edition, 2012. ISBN 978-0-12-088478-0.

[19] Thomas H. Cormen, Charles E. Leiserson, Ronald L. Rivest, and Clifford Stein. *Introduction to Algorithms*. MIT press, 3rd edition, 2009.

[20] International Organization for Standardization. *International Standard ISO/IEC 14882:2017(E) – Programming Languages – C++*. International Organization for Standardization, 2017. URL https://www.iso.org/standard/69466.html.

[21] International Organization for Standardization. *International Standard ISO/IEC*

14882:2020(E) – Programming Languages – C++. International Organization for Standardization, 2020. URL `https://www.iso.org/standard/73560.html`.

[22] Alexandre Ganea. [Clang][Driver] Re-use the calling process instead of creating a new process for the cc1 invocation. 2019. URL `https://reviews.llvm.org/D69825`.

[23] Peter Goldsborough. Emitting diagnostics in clang. URL `https://www.goldsborough.me/c++/clang/llvm/tools/2017/02/24/00-00-06-emitting_diagnostics_and_fixithints_in_clang_tools/`.

[24] Google. Google Test. 2023. URL `https://github.com/google/googletest`. C++ testing framework.

[25] International Organization for Standardization (ISO). *ISO/IEC 9899:1999 - Programming languages - C.* International Organization for Standardization (ISO), 1999. URL `https://www.iso.org/standard/23482.html`.

[26] Chris Lattner and Vikram Adve. LLVM: A Compilation Framework for Lifelong Program Analysis & Transformation. *Proceedings of the 2004 International Symposium on Code Generation and Optimization (CGO'04)*, Mar 2004.

[27] Bruno Cardoso Lopes. [RFC] Upstreaming ClangIR. January 2024. URL `https://discourse.llvm.org/t/rfc-upstreaming-clangir/76587`.

[28] Thomas J. McCabe. A complexity measure. *IEEE Transactions on Software Engineering*, SE-2(4):308–320, 1976. ISSN 0098-5589. doi: 10.1109/TSE.1976.233837.

[29] Flemming Nielson, Hanne Riis Nielson, and Chris Hankin. *Principles of Program Analysis.* Springer, Berlin, Heidelberg, 2005. ISBN 978-3-540-65410-0.

[30] Xavier Rival and Kwangkeun Yi. *Introduction to Static Analysis: An Abstract Interpretation Perspective.* The MIT Press, Cambridge, MA, USA, 2020. ISBN Your-ISBN-Number-Here.

[31] Alan M. Turing. On Computable Numbers, with an Application to the

Entscheidungsproblem. *Proceedings of the London Mathematical Society,* s2-42(1): 230–265, 1937. doi: 10.1112/plms/s2-42.1.230.

[32] Kristóf Umann. A Survey of Dataflow Analyses in Clang. October 2020. URL https://lists.llvm.org/pipermail/cfe-dev/2020-October/066937.html.

Index

www.packtpub.com

Subscribe to our online digital library for full access to over 7,000 books and videos, as well as industry leading tools to help you plan your personal development and advance your career. For more information, please visit our website.

Why subscribe?

- Spend less time learning and more time coding with practical eBooks and Videos from over 4,000 industry professionals
- Improve your learning with Skill Plans built especially for you
- Get a free eBook or video every month
- Fully searchable for easy access to vital information
- Copy and paste, print, and bookmark content

Did you know that Packt offers eBook versions of every book published, with PDF and ePub files available? You can upgrade to the eBook version at www.packtpub.com and as a print book customer, you are entitled to a discount on the eBook copy. Get in touch with us at customercare@packtpub.com for more details.

At www.packtpub.com, you can also read a collection of free technical articles, sign up for a range of free

Other Books You Might Enjoy

If you enjoyed this book, you may be interested in these other books by Packt:

LLVM Techniques, Tips, and Best Practices Clang and Middle-End Libraries

Min-Yih Hsu

ISBN: 9781838824952

- Find out how LLVM's build system works and how to reduce the building resource
- Get to grips with running custom testing with LLVM's LIT framework
- Build different types of plugins and extensions for Clang
- Customize Clang's toolchain and compiler flags
- Write LLVM passes for the new PassManager
- Discover how to inspect and modify LLVM IR
- Understand how to use LLVM's **profile-guided optimizations (PGO)** framework
- Create custom compiler sanitizers

Learn LLVM 17

Kai Nacke, Amy Kwan

ISBN: 9781837631346

- Configure, compile, and install the LLVM framework
- Understand how the LLVM source is organized
- Discover what you need to do to use LLVM in your own projects
- Explore how a compiler is structured, and implement a tiny compiler
- Generate LLVM IR for common source language constructs
- Set up an optimization pipeline and tailor it for your own needs
- Extend LLVM with transformation passes and clang tooling
- Add new machine instructions and a complete backend

Packt is searching for authors like you

If you're interested in becoming an author for Packt, please visit authors.packtpub.com and apply today. We have worked with thousands of developers and tech professionals, just like you, to help them share their insight with the global tech community. You can make a general application, apply for a specific hot topic that we are recruiting an author for, or submit your own idea.

Share your thoughts

Now you've finished *Clang Compiler Frontend*, we'd love to hear your thoughts! Scan the QR code below to go straight to the Amazon review page for this book and share your feedback or leave a review on the site that you purchased it from.

https://packt.link/r/1837630984

Your review is important to us and the tech community and will help us make sure we're delivering excellent quality content.

Download a free PDF copy of this book

Thanks for purchasing this book!

Do you like to read on the go but are unable to carry your print books everywhere? Is your eBook purchase not compatible with the device of your choice?

Don't worry, now with every Packt book you get a DRM-free PDF version of that book at no cost.

Read anywhere, any place, on any device. Search, copy, and paste code from your favorite technical books directly into your application.

The perks don't stop there, you can get exclusive access to discounts, newsletters, and great free content in your inbox daily.

Follow these simple steps to get the benefits:

1. Scan the QR code or visit the link below:

https://download.packt.com/free-ebook/9781837630981

2. Submit your proof of purchase.

3. That's it! We'll send your free PDF and other benefits to your email directly.